Networked Microgrids

Discover scalable, dependable, and intelligent solutions to the challenges of integrating complex networked microgrids with this definitive guide to the development of cutting-edge power and data systems.

It includes:

- Advanced fault management control and optimization to enable enhanced microgrid penetration without compromising reliability.
- SDN-based architectures and techniques to enable secure, reliable and fault-tolerant algorithms for resilient networked systems.
- Reachability techniques to facilitate a deeper understanding of microgrid resilience in areas with high penetration of renewables.

Combining resilient control, fast programmable networking, reachability analysis, and cyberphysical security, this is essential reading for researchers, professional engineers, and graduate students interested in creating the next generation of data-intensive self-configurable networked microgrid systems, smart communities, and smart infrastructure.

Peng Zhang is an associate professor of electrical and computer engineering at Stony Brook University.

Networked Microgrids

PENG ZHANG

Stony Brook University

CAMBRIDGE
UNIVERSITY PRESS

University Printing House, Cambridge CB2 8BS, United Kingdom

One Liberty Plaza, 20th Floor, New York, NY 10006, USA

477 Williamstown Road, Port Melbourne, VIC 3207, Australia

314-321, 3rd Floor, Plot 3, Splendor Forum, Jasola District Centre, New Delhi – 110025, India

79 Anson Road, #06–04/06, Singapore 079906

Cambridge University Press is part of the University of Cambridge.

It furthers the University's mission by disseminating knowledge in the pursuit of education, learning, and research at the highest international levels of excellence.

www.cambridge.org
Information on this title: www.cambridge.org/9781108497657
DOI: 10.1017/9781108596589

First published 2021

Printed in the United Kingdom by TJ Books Limited, Padstow Cornwall

A catalogue record for this publication is available from the British Library.

ISBN 978-1-108-49765-7 Hardback

To Haizhen, William, Henry, and Benjamin

Contents

Preface and Acknowledgments

There is an increasing demand for highly reliable and sustainable power supplies under the fast development of smart and connected communities in recent years. Meanwhile, the global trend of urbanization has been posing significant challenges on the heavily loaded and aging power infrastructures in our cities. As predicted by the United Nations, a population increase of over one billion is expected to happen in urban areas in the next 15 years. Consequently, our existing power infrastructures, which are already operating close to physical limits, may not be sustainable to support the ever-growing demand of expanding cities and smart and connected communities. As an example, there was an outage on July 13, 2019, that left tens of thousands of customers in Midtown Manhattan and the Upper West Side of New York City in blackout. Most recently, tens of thousands of homes in California experienced two rounds of planned outages in an attempt to avoid wildfires during wind events.

Microgrids have proved to be a promising paradigm of electricity resiliency. They are promising to keep local community services up and running despite utility grid outages and weather events. The vision of this book is that networking community microgrids can achieve more resilience benefits and potentially transform today's community power infrastructures into tomorrow's autonomic networks and flexible services toward self-configuration, self-healing, self-optimizing, and self-protection against cyberattacks, high levels of distributed energy resource penetration, faults, and disastrous events.

This book summarizes some of my team's initial efforts in creating smart networked microgrids by introducing new technologies that enable software-defined, hardware-independent, and resilient microgrid functions. A few students and former students assisted me in writing this book. Yanyuan Qin contributed the software-defined networking part of Chapter 4. Dr. Yifan Zhou did meticulous work to improve Chapter 5. Wenfeng Wan, with Mikhail A. Bragin and Bing Yan, made major contributions to Chapter 6. For Chapter 7, Zimin Jiang worked hard to produce most of the simulation results, and Zefan Tang helped write the majority of the chapter. Lizhi Wang developed various figures in Chapters 2 and 4. Lingyu Ren and Yanyuan Qin produced the test results in Chapter 4. Yan Li produced some figures and results in Chapter 5 and part of those in Chapters 3 and 7 along with Yanyuan Qin. Saman Dadjo Tavakoli contributed Chapter 8, where a few figures were drawn by Jiangwei Wang. Fei Feng updated some results and figures in Chapters 3 and 7. Yifan Zhou, Zefan Tang, Wenfeng Wan, Lizhi Wang, Fei Feng, Zimin Jiang, and other team members at

Stony Brook University are enthusiastically performing a series of innovative topics in microgrids and networked microgrids. We will continue to bring in novel results and tools for our community.

I am grateful to various researchers and colleagues who collaborated with me in the journey of microgrid research. The author would like to thank his former advisors Professor Yutian Liu, Professor José R. Martí, and Professor Hermann W. Dommel for their consistent support. I would like to express my gratitude to Professor Peter B. Luh, Professor Bing Wang, Professor Jun Yan, Professor Guiling Wang, Joel Rinebold, Professor Peter Willett, Professor Yaakov Bar-Shalom, Professor Chongqing Kang, Professor Yilu Liu, Professor Petar M. Djurić, and Professor Mónica F. Bugallo. Working with the colleagues has enhanced my knowledge and expanded my vision in several perspectives. I would like to thank Professor Emmanouil Anagnostou and Professor John Chandy for their leadership and support in the past years. I would like to thank Professor Matthias Althoff and Professor Weidong Xiao for discussions and their friendship. Colleagues at Eversource Energy (formerly known as Northeast Utilities) David A. Ferrante, Joseph N. Debs, Kenneth Bowes, Rodrick Kalbfleisch, Diego Castillo, Samuel Woolard, Andrew Kasznay, Camilo Serna, and Christopher Leigh deserve big thanks for their valuable support and friendship.

The National Science Foundation, Department of Energy, Brookhaven National Laboratory, and Eversource Energy generously supported my research, making possible the microgrid research reported in this book. In particular, my sincere thanks go to Dr. Radhakishan Baheti, Dr. Eyad Abed, Dr. Meng Yue, Robert J. Lofaro, and Dr. Ali Ghassemian for their great support and encouragement over the years.

My sincere gratitude goes to Emma Burris-Janssen, who proofread and edited the draft throughout the writing process. Many thanks to Julia Ford, Sarah Strange, Elizabeth Horne, and the entire Cambridge University Press editorial team for their excellent work and support.

Once again, I thank all those who helped, supported, and collaborated with me in the journey of power systems research.

Notation

AC	alternating current
AFM	active fault management
AMI	advanced metering infrastructure
ComPF	compositional power flow
DAEs	differential-algebraic system of equations
DAPI	distributed averaging proportional-integral
DA-SLR	distributed and asynchronous surrogate Lagrangian relaxation
DC	direct current
DER	distributed energy resource
DFA	distributed formal analysis
DOE	Department of Energy
DQG	quasidiagonalized Geršgorin (DQG) theory
D-STATCOM	Distribution Static Synchronous Compensator
EMF	electromotive force
EMS	energy management system
FA	formal analysis
FRT	fault ride-through
HIL	hardware-in-the-loop
HTS	host tracking service
ID	identification
IoT	internet of Things
IP	internet protocol
LC	local controller
LFC	load frequency control
MAC	media access control
MIMO	multi-input multi-output
MIP	mixed-integer programming
microPMU	microphasor measurement unit
microRAS	microgrid remedial action scheme
NMs	networked microgrids
NMCC	networked microgrids coordination center
NSF	National Science Foundation
ODE	ordinary differential equation

OPF	optimal power flow
PAC	protection, automation, and control
PCC	point of common coupling
PLL	phase lock loop
PV	photovoltaic
PWM	pulse-width modulation
QoS	quality of service
QR	eigenvalue algorithm through QR decomposition
RGA	relative gain array
RTU	remote terminal unit
SAIDI	system average interruption duration index
SAIFI	system average interruption frequency index
S&CC	smart and connected community
SDASD	software-defined active synchronous detection
SDN	software-defined networking
SD^2N	software-defined distribution network
SPM	smart programmable microgrid
SQP	sequential quadratic programming
SVD	singular value decomposition
TCP	transmission control protocol
UDN	urban distribution network
UDP	user datagram protocol
VM	virtual machine
VSG	virtual synchronous generator
VSI	voltage source inverter

Part I

Fundamentals

1 Introduction

1.1 Empowering Smart and Connected Communities through Microgrids and Networked Microgrids

The keystone of smart and connected communities (S&CCs) is a resilient electric network capable of supporting critical infrastructures such as water, food, public safety, transportation, waste management, communication, and other functions that are vital for citizens [1].

Existing electric networks, however, cannot sustain growing communities' ever-increasing demands. According to the Department of Energy, US customers experienced power outages for an average of 198 minutes in 2015. One state even lost power for over 14 hours. Over the same period of time, customers in Japan and Korea only lost power for an average of less than 10 minutes and less than 12 minutes, respectively [2]. Furthermore, distributed energy resources (DERs), such as intermittent photovoltaics (PVs) increasingly installed in the United States communities, fail to improve electricity resilience, because they cannot ride through sustained grid contingencies. In addition to these challenges, extreme weather events and cyberattacks [3] can lead to catastrophic blackouts. Actually, in the United States, thousands of major blackouts have occurred in the past three decades causing over $1 trillion in damages and enormous social upheavals [4]. Of these outages, over 90% occurred along electric distribution systems under extreme weather events [5]. In light of this, academia [6, 7], industry [8, 9], and government [10] now share a strong consensus that enhancing distribution systems' resilience is an important focus of research [11, 12].

Microgrids offer a promising paradigm for increasing electricity resiliency for customers [42]. Normally, a microgrid is still a local power distribution grid, but it is an autonomous system rather than a traditional passive network. It is often designed to provide a more reliable and resilient electrical and heat energy supply for a community such as a commercial building, a residential area, a military base, a university campus, or a mix of these. A salient feature of a microgrid is that it operates not only when it is coupled with the main grid (grid-connected mode) but also when it is disconnected from the main grid during emergencies (islanded mode). Though microgrids are more similar than different, they can be defined in a variety of ways, a trend that is likely to increase with advancements in microgrid technology. As an example, the International Council on Large Electric Systems (CIGRE) C6.22 Working Group [13] has defined microgrids as follows:

Microgrids are electricity distribution systems containing loads and distributed energy resources, (such as distributed generators, storage devices, or controllable loads) that can be operated in a controlled, coordinated way either while connected to the main power network or while islanded.

In theory, the microgrid model provides a potent option for preventing power outages. Because microgrids can operate autonomously, accommodate renewable resources, and remain immune to weather damage, they promise to prevent local power outages. For example, the author's research [14, 15] has quantified the resilience benefits of microgrids for hardening critical infrastructures within 15 Connecticut cities/towns under various weather conditions (e.g. major storm; tropical storm; Categories 1, 2, and 3 storms). The resiliency and reliability of microgrids are also being proved in the real world. For instance, during the 2017 flooding of Hurricane Harvey in the Houston area, a handful of microgrids allowed the Texas Medical Center, as well as some local gas stations and stores, to continue operating despite severe utility grid outages [16].

The need for highly resilient and clean energy resources in important load centers calls for the adoption of microgrids in cities and smaller communities. For instance, large data centers, which are critical for the United States' growing digital economy, suffered from an average of $2.4 million in power outage costs per data center in 2015, an 81% increase compared to 2010 [17, 18]. Thus, in the coming decade, data centers are expected to represent up to 40% or more of the total microgrid market [19]. As the world's population becomes increasingly concentrated in urban centers, there will be a corresponding rise in the demand for electricity, a trend that will pose significant challenges to cities' aging power infrastructures. The power outage that occurred on July 13, 2019, in Midtown Manhattan and New York City's Upper West Side was a typical example of the daunting problems faced by America's aging electrical infrastructure. The author believes that, in order to achieve an adequate level of electrical resiliency, microgrids will be increasingly accepted and installed in cities or S&CCs with dense populations and critical loads. Although individual microgrids can surely provide electricity resilience to their own customers, they seldom are able to improve the reliability and resilience of the main grid [20]. On the other hand, our preliminary research has shown [20] that coordinated *networked microgrids* can potentially help restore neighboring distribution grids after a major blackout. Networked microgrids also promise to significantly enhance day-to-day reliability and performance by improving both the System Average Interruption Duration Index (SAIDI) and the System Average Interruption Frequency Index (SAIFI). Actually, as anticipated by the US Department of Energy, the research and development (R&D) of networked microgrids will lead to the next wave of smart grid research, which will help achieve the nation's grid modernization vision toward climate adaptation and resiliency [21].

1.2   Challenges in Networked Microgrids

Networked microgrids – also referred to as coupling microgrids – consist of an interconnected cluster of microgrids in close electrical or spatial proximity to allow both coordinated energy management and interactive support and exchange [21–23].

Recently, studies have begun to investigate how power interchange among networked microgrids can help support more customers during situations of islanding or reduce power deficiencies in individual microgrids caused by intermittent renewables [22]. Other studies have proposed that decentralized dispatch be used to ensure the steady-state operation of networked microgrids [24]. Although promising results have been achieved for steady-state operations, dynamic performance under grid disturbances and faults remains an intractable challenge that compromises microgrids' resiliency, thus preventing the wide adoption of networked microgrids. Following is a summary of the three main problems that must be addressed before networked microgrids can be adopted more widely:

- *Understanding the dynamics of networked microgrids.* There is a methodological gap in computing the security indicators – including stability margin/region – for networked microgrids [25]. This challenge originates from the salient features of microgrids such as uncertainties, nonsynchronism, fast ramp rates, and low inertia. Therefore, a compelling question is this: *How can microgrid stability be assessed and enhanced for the improvement of networked microgrids' situational awareness and controllability? This question is key to determining whether networked microgrids can be used as dependable resiliency resources [26].*
- *Impact of high levels of microgrid penetration.* Although a high level of microgrid penetration is necessary to provide adequate capacity for preventing and mitigating grid outages [27], it creates a difficult dilemma. On the one hand, microgrids are designed to trip offline (or have their grid-interface converters change to zero power mode) and operate in islanding mode during grid contingencies. As a result, when many microgrids are interconnected, a grid disturbance can induce the sudden loss of a large number of microgrids, significantly increasing the risk of blackout in both distribution and transmission grids (see an ISO New England report [28]). On the other hand, forcibly coupling a network of microgrids with a disturbed main grid can lead to catastrophic mutual impacts, such as excessive fault current contributions into the main grid as well as large disturbances propagating into and out of the microgrids. This raises a fundamental question: *How can the fault responses of networked microgrids be managed so that they can provide critical ancillary support as required by the new Institute of Electrical and Electronics Engineers (IEEE) Standards 1547 and 2030 [29] without causing excessive (or inadequate) fault current magnitudes in the main grid, as well as internal and external instabilities?*
- *Bottleneck in the communication infrastructure.* Resiliency issues such as time delay, congestion, failures, and cyberattacks can significantly compromise the functionality of microgrids' communication networks, which are indispensable to microgrid stability control and fault management. Moreover, traditional cyberarchitecture has been found unable to adapt to the ever-increasing pace of functional and structural changes in microgrids and distribution grids [30, 31]. An essential problem is *how to design an innovative communication architecture to enable ultrafast microgrid control, respond to and mitigate threats and emergencies, and fully support scalable networked microgrids.*

This book aims to close the knowledge gap highlighted in the preceding so that reliable networked microgrids can boost distribution grid resiliency. The networked microgrid technologies discussed in this book are based on the author's study of microgrids and networked microgrids since 2010. In particular, this book summarizes several new findings from our recent research projects, including the following:

- EAGER: US Ignite: Enabling Highly Resilient and Efficient Microgrids through Ultra-Fast Programmable Networks, US National Science Foundation
- US Ignite: Focus Area 1: SD^{2}N: Software-Defined Urban Distribution Network for Smart Cities, US National Science Foundation
- Enabling Reliable Networked Microgrids for Distribution Grid Resiliency, US National Science Foundation
- SCC: Empowering Smart and Connected Communities through Programmable Community Microgrids, US National Science Foundation
- Formal Analysis for Dynamic Stability Assessment of Large Interconnected Grids under Uncertainties, US Department of Energy
- Academic Plan Level I: Software Defined Smart Grid, University of Connecticut

This book has three main objectives:

- *To establish a formal analysis method to tractably assess networked microgrid stability.* Modeling and analysis of uncertainties due to high levels of renewable generation will be addressed. A formal theory will be developed to increase situational awareness and unlock the potential of networked microgrids as primary resilience resources.
- *To devise a new concept of microgrid active fault management (AFM) through online distributed optimization.* The new approach allows networked microgrid riding through balanced and unbalanced grid disturbances, systematically enhancing grid resiliency rather than negatively impacting the disturbed grid. In the future, formal methods will be used to ensure provably correct AFM with guaranteed performance.
- *To build a software-defined networking (SDN) architecture to enable highly resilient networked microgrids.* As a new paradigm for computer networking, SDN provides run-time programmability and unprecedented flexibility in managing communication networks. SDN-based algorithms and a coordinated framework will be investigated to address communication challenges for networked microgrids, including time delays, network failures, service support quality, and cyberattacks.

As one of the first monographs on networked microgrids, this book is expected to serve as a reference for academia and industry. The contents will be useful for professors who plan to offer courses related to microgrids, active power distribution, and distributed energy resources. The topics discussed in this book will include power systems, power electronics, dynamics and stability, control theory, computer networking, and cybersecurity. Thus, readers who are inquisitive and curious about the integration of those techniques in the vibrant field of microgrids will find this book a useful introductory textbook. Reachability techniques will be devised to facilitate a deeper understanding of microgrid resilience under renewable energy's high penetration levels. The idea of an AFM-oriented power electronic control with distributed optimization will pave the way for a high penetration of microgrids without compromising grid reliability. The novel SDN-based architecture and techniques will open the door for innovations in devising secure, reliable, and fault-tolerant algorithms for managing resilient networked systems such as multiple microgrids and active distribution networks. Overall, the proposed new model-based and data-intensive technologies together will provide scalable, dependable, and intelligent solutions to traditionally intractable problems in integrating complex networked microgrids.

1.3 Overview of Topics

This book focuses on establishing several theoretical foundations for reliable networked microgrid operations and analysis in the face of various cyber and physical disturbances. It leverages our recent research in resilient microgrid control, ultrafast programmable networking, dynamical systems, cyberphysical security, and real-time hardware-in-the-loop microgrid testbeds to resolve the previously intractable problems posed by integrating scalable microgrids with a high penetration of renewable energy resources. The book includes the following chapters:

- Chapter 1. "Introduction"
  This chapter provides fundamental information about both microgrids and networked microgrids. It also previews the topics discussed in the rest of the book.
- Chapter 2. "Basics of Microgrid Control"
  This chapter covers the basics of hierarchical control for microgrids.
- Chapter 4. "Compositional Power Flow for Networked Microgrids"
  This chapter introduces a privacy-preserving, distributed power flow approach to networked microgrids' situational awareness.
- Chapter 4. "Resilient Networked Microgrids through Software-Defined Networking"
  This chapter discusses an SDN-enabled architecture that transforms isolated local microgrids into integrated networked microgrids.
- Chapter 5. "Formal Analysis of Networked Microgrids' Dynamics"
  This chapter establishes an innovative and tractable method called distributed formal analysis (DFA) for assessing the stability of networked grids under uncertainties.
- Chapter 6. "Active Fault Management for Networked Microgrids"
  This chapter devises a concept of AFM enabled through online distributed optimization.
- Chapter 7. "Cyberattack-Resilient Networked Microgrids"
  This chapter aims to understand, model, and mitigate the cybersecurity risks to the networked microgrids' operations.

- Chapter 8. "Networked DC Microgrids"
  This chapter focuses on a dynamic modeling and stability assessment of bipolar direct current (DC) microgrids and gives an introduction to the stability of networked DC microgrids.
- Chapter 9. "Software-Defined Distribution Network"
  This chapter introduces a future software-defined, hardware-independent microgrid infrastructure for smart and connected communities.
- Chapter 10. "Future Perspectives: Programmable Microgrids"
  This chapter provides the author's visions for next-generation microgrid research and development.

The book adopts a cyber-physical system (CPS) approach, presenting a rigorous unification of the theoretical underpinnings behind the emerging field of networked microgrids. It draws on a diverse set of cyber-physical system methods to tackle open challenges in the operation and analysis of networked microgrids. It includes a power flow algorithm for islanded networked microgrids without any slack buses, formal methods via reachable set calculations for evaluating networked microgrids stability regions, online distributed optimization for microgrid inverters to ride through faults, active defense strategies to detect and mitigate deception attacks on microgrids, stability of bipolar DC microgrids, software-defined networking solutions to network individual microgrids, and resilience enhancement for physical and cybernetworks.

The book is suitable for classroom use or as a reference for professionals. It consists of accurate modeling details for microgrid components as well as test cases for networked microgrids. The book can be used for a semester-long course aimed at senior undergraduates or graduate students in electrical and computer engineering. It can be used as training material for continuing education purposes (e.g., power engineering graduate certificate programs) for electrical engineers. Power industry professionals will find the algorithms and architectures provided in this book useful for solving some challenges in their job duties. Based on the models provided in this book, researchers and professionals will be able to integrate our solutions to existing commercial tools and develop new tools. In general, this book can serve as a guide for industry and military professionals to understand the principles of microgrid operations and provide valuable tools for the planning, design, operation, and protection of microgrids.

References

[1] M. La Scala, S. Bruno, C. A. Nucci, S. Lamonaca, and U. Stecchi, *From Smart Grids to Smart Cities: New Challenges in Optimizing Energy Grids*. John Wiley & Sons, 2017.

[2] Department of Energy, "Transforming the Nation's Electricity System: The Second Installment of the Quadrennial Energy Review," Jan. 2017.

[3] T. Flick and J. Morehouse, *Securing the Smart Grid: Next Generation Power Grid Security*. Elsevier, 2010.

[4] D. T. Ton and W. T. Wang, "A More Resilient Grid: The US Department of Energy Joins with Stakeholders in an R&D Plan," *IEEE Power and Energy Magazine*, vol. 13, no. 3, pp. 26–34, 2015.

[5] R. J. Campell, "Weather-Related Power Outages and Electric System Resiliency," Washington, DC: Congressional Research Service, 2012. https://fas.org/sgp/crs/misc/R42696.pdf.

[6] S. Martin, G. Deffuant, and J. M. Calabrese, "Defining Resilience Mathematically: From Attractors to Viability," in *Viability and Resilience of Complex Systems*, G. Deffuant and N. Gilbert, eds. Springer, 2011, pp. 15–36.

[7] C. Folke, S. R. Carpenter, B. Walker, M. Scheffer, T. Chapin, and J. Rockström, "Resilience Thinking: Integrating Resilience, Adaptability and Transformability," *Ecology and Society*, vol. 15, no. 4, p. 20, 2010.

[8] A. Rose, "Economic Resilience and Its Contribution to the Sustainability of Cities," in *Resilience and Sustainability in Relation to Natural Disasters: A Challenge for Future Cities*, P. Gasparini, G. Manfredi, and D. Aprone, eds. Springer, 2014, pp. 1–11.

[9] L. Molyneaux, L. Wagner, C. Froome, and J. Foster, "Resilience and Electricity Systems: A Comparative Analysis," *Energy Policy*, vol. 47, pp. 188–201, 2012.

[10] J. Inglis, S. Whittaker, A. Dimitriadis, and S. Pillora, "Climate Adaptation Manual for Local Government: Embedding Resilience to Climate Change," Australian Centre of Excellence for Local Government, University of Technology, Sydney. https://opus.lib.uts.edu.au/handle/10453/42120, 2014.

[11] M. McGranaghan, M. Olearczyk, and C. Gellings, "Enhancing Distribution Resiliency: Opportunities for Applying Innovative Technologies," *Electricity Today*, vol. 28, no. 1, pp. 46–48, 2013.

[12] J. Carlson, R. Haffenden, G. Bassett, W. Buehring, M. Collins III, S. Folga, F. Petit, J. Phillips, D. Verner, and R. Whitfield, "Resilience: Theory and Application." Argonne National Laboratory (ANL), Tech. Rep., 2012.

[13] C. Marnay, S. Chatzivasileiadis, C. Abbey, R. Iravani, G. Joos, P. Lombardi, P. Mancarella, and J. von Appen, "Microgrid Evolution Roadmap," in *Smart Electric Distribution Systems and Technologies (EDST), 2015 International Symposium on*. IEEE, 2015, pp. 139–144.

[14] P. Zhang, G. Li, and P. B. Luh, "Reliability Evaluation of Selective Hardening Options," State of Connecticut, Tech. Rep., May 2015.

[15] G. Li, P. Zhang, P. B. Luh, W. Li, Z. Bie, C. Serna, and Z. Zhao, "Risk Analysis for Distribution Systems in the Northeast U.S. under Wind Storms," *IEEE Transactions on Power Systems*, vol. 29, no. 2, pp. 889–898, March 2014.

[16] Harvey's Devastation Shows the Need for Distributed Energy, Microgrids during Disasters. [Online]. Available: www.greentechmedia.com/articles/read/harveys-devastation-shows-the-need-for-distributed-energy-microgrids-during.

[17] Ponemon Institute and Emerson Network Power, "Cost of Data Center Outages," Ponemon Institute, Tech. Rep, Tech. Rep., 2016.

[18] Eaton, "Blackout and Power Outage Tracker," http://powerquality.eaton.com/blackouttracker/.

[19] Navigant Research, "Data Centers and Advanced Microgrids," Navigant Research, Tech. Rep., 2017.

[20] Z. Bie, P. Zhang, G. Li, B. Hua, M. Meehan, and X. Wang, "Reliability Evaluation of Active Distribution Systems Including Microgrids," *IEEE Transactions on Power Systems*, vol. 27, no. 4, pp. 2342–2350, November 2012.

[21] K. Cheung, "DOE Perspective on Microgrids," keynote presentation, 2015 Applied Power Electronics Conference and Exposition, March 2015.

[22] F. Shahnia, S. Bourbour, and A. Ghosh, "Coupling Neighboring Microgrids for Overload Management Based on Dynamic Multicriteria Decision-Making," *IEEE Transactions on Smart Grid*, vol. PP, no. 99, pp. 1–1, 2015.

[23] Z. Wang, B. Chen, J. Wang, M. M. Begovic, and C. Chen, "Coordinated Energy Management of Networked Microgrids in Distribution Systems," *IEEE Transactions on Smart Grid*, vol. 6, no. 1, pp. 45–53, 2015.

[24] Z. Wang, B. Chen, J. Wang, and J. Kim, "Decentralized Energy Management System for Networked Microgrids in Grid-Connected and Islanded Modes," *IEEE Transactions on Smart Grid*, vol. PP, no. 99, pp. 1–1, 2015.

[25] G. Strbac, N. Hatziargyriou, J. Carvalho Lopes, C. Moreira, A. Dimeas, and D. Papadaskalopoulos, "Microgrids: Enhancing the Resilience of the European Megagrid," *IEEE Power and Energy Magazine*, vol. 13, no. 3, pp. 35–43, May 2015.

[26] Presidential Policy Directive, "Critical Infrastructure Security and Resilience," *Department of Homeland Security Std*, 2013.

[27] J. Li, X.-Y. Ma, C.-C. Liu, and K. P. Schneider, "Distribution System Restoration with Microgrids Using Spanning Tree Search," *IEEE Transactions on Power Systems*, vol. 29, no. 6, pp. 3021–3029, 2014.

[28] M. Henderson, "Impacts of Transmission System Contingencies on Distributed Generation – Overview." DG Forecast Working Group Meeting, ISO New England, Tech. Rep., December 2013.

[29] T. S. Basso, *IEEE 1547 and 2030 Standards for Distributed Energy Resources Interconnection and Interoperability with the Electricity Grid*, 2014.

[30] M. Tahir and S. K. Mazumder, "Self-Triggered Communication Enabled Control of Distributed Generation in Microgrids," *IEEE Transactions on Industrial Informatics*, vol. 11, no. 2, pp. 441–449, 2015.

[31] L. Ren, Y. Qin, Y. Li, P. Zhang, B. Wang, P. B. Luh, S. Han, T. Orekan, and T. Gong, "Enabling Resilient Distributed Power Sharing in Networked Microgrids Through Software Defined Networking," *Applied Energy*, vol. 210, pp. 1251–1265, 2018.

2 Basics of Microgrid Control

2.1 Microgrid Operation

Today's microgrids normally contain the following components: a localized group of electricity generators, including solar, wind, fuel cell, combined heat and power, and diesel generators; loads such as buildings, factories, and shops; and storage, which may be coordinated and managed by a microgrid energy management system (EMS) or a platform with similar functionalities (see Figure 2.1). Because of their multiple dispersed energy resources and their ability to isolate from the main grid, microgrids provide a promising paradigm for improving the resiliency of the electric distribution infrastructure.

As an illustrative example, Canada-based Alectra Utilities (formerly PowerStream) has created a small-scale microgrid demonstration project at its head office to validate the potential economic and resilience benefits to their customers. As can be seen in Figure 2.2, the Alectra microgrid consists of a solar array, a wind turbine, a natural-gas generator, a lead-acid battery, and a lithium-ion battery. The microgrid loads include the lighting, air conditioning, and refrigeration for the company's cafeteria, entrance area, and electric vehicle charging station.

A microgrid may operate in four modes: grid-connected (i.e., connected to the main grid), islanded (i.e., disconnected from the main grid), islanding (i.e., transitioning from grid-connected to islanded), or reconnected (i.e., transitioning from islanded to grid-connected). When normal conditions apply to the main grid, islanding can be conducted intentionally, but it can also be performed unintentionally when the grid is subjected to abnormal conditions (e.g., faults). In this book, we focus on microgrids with a high penetration of renewable energy resources. Though these microgrids bring significant environmental benefits, they also pose major challenges in management and control. For instance, unintentional islanding, also known as an emergency operation, is particularly challenging for such microgrids since renewable energy resources are intermittent, uncertain, and have much smaller inertias than traditional power generation plants [1, 2]. In such microgrids, it is extremely important to achieve fast voltage and frequency control in an emergency operation; otherwise, the system may lose its balance between load and generation, which could lead to eventual collapse.

Like many functionalities, rapid voltage and frequency control rely on the microgrid's communication infrastructure (see Figure 2.1b). Communication can be conducted from the microgrid's EMS to the main grid's EMS, between the microgrid's

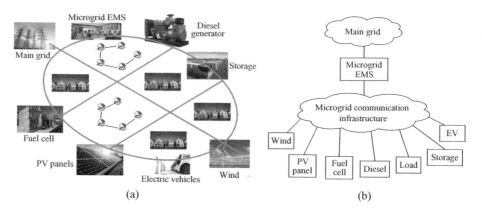

Figure 2.1 Illustration of a microgrid. (a) The various components in a microgrid. (b) Schematic view of a microgrid where the various components are connected by a communication infrastructure. Figure courtesy of Professor Bing Wang

Figure 2.2 A microgrid control room at Alectra Utilities (formerly PowerStream).

EMS and its various components, and among the various components. There are many types of communication data, from small periodic control messages to large data messages, with diverse quality of service (QoS) requirements depending on the type of data and the microgrid's mode. In the most stringent scenarios (e.g., microgrid control messages when entering the emergency operation mode), the delay requirement may be just a few milliseconds. In less stringent scenarios (e.g., during steady-state control, intentional islanding, and reconnection), the delay requirement may be as long as a few seconds. The least stringent scenarios apply to the transferral of energy management information, which can tolerate delays as long as a few minutes.

As mentioned earlier, the networking technologies used in microgrid deployments may vary depending on a number of factors such as deployment costs, required coverage, and suitability for control. For instance, when deployment costs are not a concern, adopting dedicated networks of redundant fiber optics can provide fast and reliable communication; in areas with well-established wired infrastructures (e.g., Ethernet), adopting the existing infrastructure would be a natural choice. However, in certain rural or disaster areas, it might be too costly or infeasible to establish a wired network, making wireless technologies (e.g., WiFi, WiMax, or cellular network) the only option. No matter what networking technologies are used, the communication infrastructure must satisfy the diverse QoS requirements and be resilient to network failure (in the form of degraded performance or complete failure) caused by congestion, interference, or hardware/software faults [3]. The latter requirements demand that the communication infrastructure provide multiple network paths, through rich connectivity in network topology or multiple networking technologies (e.g., Ethernet, WiFi, WiMax, cellular networks) [4]. Because these technologies must be managed carefully to ensure that a microgrid can operate properly in the face of a network failure, a software-defined network should be used. As detailed in Chapter 4, this general communication architecture can be used in a wide range of microgrid deployments, despite the underlying networking technologies that are being adopted.

2.2 Microgrid Control

With the support of the communication infrastructure, microgrid control schemes – especially those designed for the operation of microgrids under islanded situations – can be developed. Microgrid control is particularly challenging when there is a high penetration of renewable energy resources in the microgrid. In this chapter, we focus on droop control, an important primary control strategy for rapidly restoring a microgrid's voltage and frequency. Secondary control, remedial action scheme (microRAS), and optimal power flow will also be discussed briefly.

2.2.1 Hierarchical Control Principle

The popular droop control scheme implemented in microgrids [5–16] was perhaps inspired by the droop control of synchronous generators in a traditional power system [17, 18]. In traditional, vertically integrated utilities, the power system control follows a hierarchical control framework that includes the primary control, the secondary control, and the tertiary control. Even though there has been a trend of utility restructuring and deregulation, the hierarchical control framework has been largely retained [19]. The primary control, or droop control, is at the bottom level and is achieved through the turbine governor and its speed-droop characteristic; all the control commands from the upper levels are executed at this level. The droop control is a decentralized proportional control, where only local measurements are used as feedback signals and the droop coefficient is the control gain [20].

The secondary control is used to force the droop control to eliminate frequency deviations and maintain the agreed tie line power flows between utilities. In isolated grids, the secondary control is degenerated into the frequency control only [18]. Because the secondary control normally relies on communication networks, it is slower than the primary control. The tertiary control is slower than the secondary control, and at the EMS level, it may encompass more advanced functions such as optimal power flow (OPF) and economic dispatch [21].

In a traditional power system, one or more power plants are enrolled in the load frequency control (LFC) to maintain frequency. In general, system frequency is governed by load and generation conditions. If the load suddenly increases and becomes greater than the power generation, the frequency tends to decrease. Accordingly, the increase in load will be supplied by the kinetic energy released from the generators [22]. Once the turbine governors of all the power plants sense the frequency drop, the droop control starts to increase power generation by the required amount, resulting in an improved but reduced frequency determined by the speed-droop curves. Later, the LFC will further increase power generation in the generators registered on the LFC to bring the system frequency back to the scheduled or nominal value, which is the secondary control. Similarly, if the load suddenly decreases and becomes lower than the power generation, the excessive generation will be converted to kinetic energy, which speeds up all the generators until the frequency increase is sensed by the governors [22]. The droop control and secondary control will then be initiated to recover the frequency.

In a turbine governor system, a speed sensor (e.g., a flyball for a mechanical-hydraulic governor and a frequency transducer for an electrohydraulic governor [23]) measures the speed of the turbine and opens the steam valves on the steam turbines or the wicket gates on the hydroturbines if the speed decreases [22]. This turbine control system produces a steady-state speed-droop curve as shown in Figure 2.3. Thus, the droop control is an inherent characteristic of a turbine generator system.

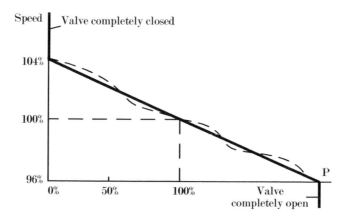

Figure 2.3 Steady-state speed droop of a synchronous generator. Figure courtesy of Dr. H. W. Dommel's lecture notes [22]

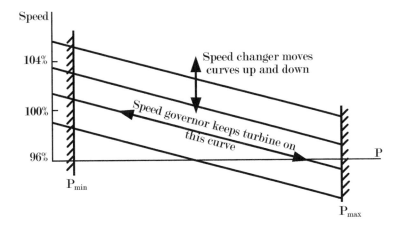

Figure 2.4 Secondary control of a synchronous generator. Figure courtesy of Dr. H. W. Dommel's lecture notes [22]

For a grid-tied synchronous generator without a secondary control device, the turbine power at the nominal speed will always be the rated power (see Figure 2.3). To change power output at will, a speed changer is equipped to change the valve's or the wicket gate's position independently of the primary control action, which can create a series of steady-state speed-droop curves as shown in Figure 2.4. Thus the power output at the nominal speed can be set to any value within P_{min} and P_{max} through the speed changer [22], which enables the secondary control for the LFC generators to fully recover the system frequency.

2.2.2 Droop Control for Microgrids

A majority of the renewable distributed energy resources (DERs) in microgrids produce direct current (DC) power (e.g., photovoltaics) or operate at a high frequency (e.g., single-shaft microturbines) or a variable frequency (e.g., wind turbine generators). Those DERs are interconnected with the alternating current (AC) microgrid backbone through inverters [24]. As a result, the droop control of inverter-dominated microgrids is deployed on DER inverters. Unlike the speed governor for a set of synchronous generators described in Subsection 2.2.1, inverters do not have any inherent tendency toward the speed-droop characteristic, meaning a droop effect must be artificially created through the inverter controller.

One possible motivation for using a conventional droop control strategy at the microgrid level is to help the main grid and its microgrids respond coherently to frequency changes, which in turn helps achieve a consistent control effect throughout the grid that can reduce concerns about microgrids' impact on grid stability and reliability [25]. In addition, droop control can be achieved without a centralized supervisory control, as it uses local variables to regulate power output. Thus, it is a modular and plug-and-play design in itself.

Droop control can be deployed on DER inverters operating in a grid-forming mode, i.e., voltage source inverters (VSIs). Here, a grid-forming inverter means an inverter that is controlled such that its output voltage can be specified as needed [20]. To achieve droop control, a cascaded control scheme including an outer voltage loop and an inner current loop can be integrated in a VSI [20]. For one or more DERs participating in microgrid stabilization and voltage recovery, frequency-droop and voltage-droop control strategies are used to share active and reactive power among DERs. One straightforward approach to implementing droop control is to use P as a function of f and Q as a function of V, resulting in the so-called $f - P/V - Q$ control, expressed as follows:

$$P = P^* + K_f(f^* - f) \tag{2.1a}$$

$$Q = Q^* + K_V(V^* - V). \tag{2.1b}$$

Some of the literature on this subject [25] argues that measuring instantaneous real power is more viable than obtaining an accurate measurement of instantaneous frequency. This might have been a reason why a droop control with f as a function of P and V as a function of Q was proposed, where the VSI output power is measured and used to regulate its output frequency and voltage. This so-called $P - f/Q - V$ control is expressed as follows:

$$f = f^* + K_P(P^* - P) \tag{2.2a}$$

$$V = V^* + K_Q(Q^* - Q). \tag{2.2b}$$

Through the aforementioned $P - f/Q - V$ droop control, microgrid inverters are able to mimic the behavior of a synchronous generator that increases its active power output in response to a load increase and reduces its frequency, as Figure 2.3 illustrates. Here, f^* and V^* are the reference values for the microgrid's frequency and the amplitude of the inverter output voltage, respectively, and P^* and Q^* are the corresponding active and reactive power outputs. One type of typical droop control scheme deployed on a two-level three-phase inverter is illustrated in Figure 2.5.

The inner current-controlled loop determines the reference voltage waveforms for the pulse-width modulation (PWM) of the VSI [26]. The control scheme can be implemented in a synchronous $dq0$ frame through which the three-phase output currents are transformed into their direct (d-axis) and quadrature (q-axis) components i_d and i_q. The d-axis and q-axis currents pass through a low-pass filter. Then they are compared with reference signals i_{dref} and i_{qref}, which are specified by the outer voltage control loop. The error signals are applied to a proportional-integral (PI) control block with a current limiter, which determines the d- and q-components of reference voltages v_{dref} and v_{qref} after including the voltage feed-forward terms and the cross-coupling elimination terms. Eventually, the three-phase reference signals for the PWM signal generator are specified by transforming the dq quantities to abc quantities.

The outer loop realizes the $P - f/Q - V$ control purpose, where the VSI active power output is used to regulate the voltage angle through the integration of the frequency, and the VSI reactive power output is used to regulate the voltage amplitude [27]. The controller's output signals are v_{dref} and v_{qref}, which will be used to generate

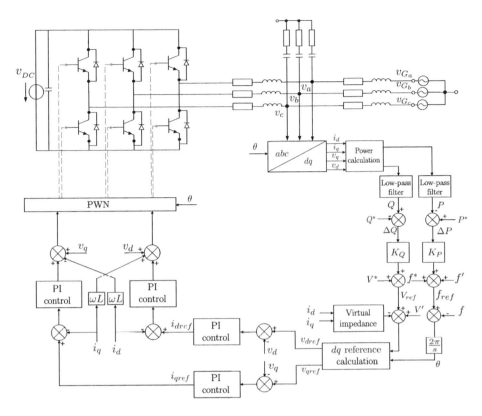

Figure 2.5 A typical droop control scheme implemented on a three-phase inverter.

the current references for the inner loop (see Figure 2.5). It should be noted that the droop control expressed in (2.2) is based on a heuristic correlation between $P - f$ and $Q - V$ under the assumption of an inductively dominant grid (backbone feeders with small R/X ratios). For microgrids where the inductive assumption no longer holds, however, the standard droop control in (2.2) may need to be modified to enable better performances [6, 16, 28]. Furthermore, the droop control can also include a virtual impedance loop (see Figure 2.5) to achieve extra desirable properties. Readers are referred to references such as [5, 29] for details.

The *secondary control* aims to fully restore the microgrid's voltage and frequency by eliminating the remaining deviations after the droop control is applied. To achieve the secondary control, the microgrid's frequency and the terminal voltage of a DER participating in the secondary control are compared with the reference frequency and voltage amplitude f^* and V^*, respectively [30]. The error signals are regulated through the secondary controller to generate secondary control signals as follows:

$$f' = K_{Pf}(f^* - f) + K_{If} \int (f^* - f) + \Delta f_s \qquad (2.3a)$$

$$V' = K_{PV}(V^* - V) + K_{IV} \int (V^* - V), \qquad (2.3b)$$

where K_{Pf}, K_{If}, K_{PV}, and K_{IV} are the secondary control parameters, and Δf_s is a synchronization term that facilitates the microgrid's synchronization with the main grid. When the microgrid is islanded, Δf_s is always zero.

As can be seen in Figure 2.5, signals f' and V' are then added to the droop control to push the droop characteristics of each DER up or down to restore the frequency and the voltage to their nominal values. The secondary control in (2.3) can be implemented by a centralized controller or by a distributed controller through distributed averaging techniques [31].

2.2.3   Master–Slave Control

An islanded microgrid can also adopt the so-called master–slave control [11], where a leader DER or an energy storage unit operates in the grid-forming mode to provide voltage and frequency references for the microgrid (V/f control) while the rest of the DERs operate in the grid-following mode (PQ control) [27]. The "master" DER, therefore, should have sufficient capacity to absorb load variations during islanded operations. A typical V/f control scheme and a typical PG control scheme deployed on a two-level, three-phase inverter are illustrated in Figures 2.6 and 2.7, respectively.

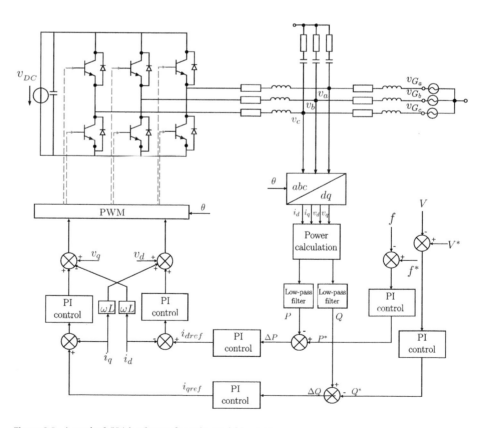

Figure 2.6 A typical V/f scheme for microgrid inverter.

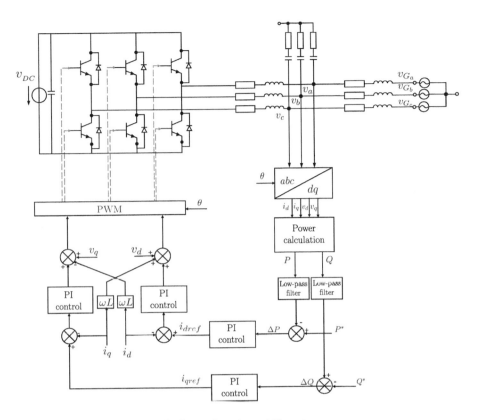

Figure 2.7 A typical PQ control scheme for microgrid inverter.

For the V/f control in Figure 2.6, the inner current-controlled loop can be similar to that described in Subsection 2.2.2. The outer loop aims to provide reference frequency and voltage amplitude at the DER's coupling point, somehow emulating the behavior of a slack bus. Measured frequency f is compared with its reference f_{ref}, and the error is regulated by a PI controller to provide reference signal P_{ref} for active power. Measured voltage V is compared with its reference V_{ref}, and the error is regulated by a PI controller to provide reference signal Q_{ref} for reactive power [27]. The two references are then used to generate reference signals for the inner loop. As a result, a DER's V/f control will shift its active and reactive power in such a way that its terminal voltage and frequency will remain relatively constant.

A major difference between the PQ control in Figure 2.7 and a grid-forming control is that the outer loop of the PQ control is a power control instead of a voltage-controlled one. A DER's active and reactive power outputs can be calculated, filtered, and then compared with their reference values (see Figure 2.7). The errors in power outputs are then regulated through a PI controller to obtain current references i_{dref} and i_{qref}, which will be the input signals for the inner control loop. The inner current-controlled loop for the PQ control can be similar to the one described in Subsection 2.2.2. In addition to the aforementioned standard PQ control scheme, many variations can be derived to achieve effective or simplified PQ control for different use cases.

2.2.4   Tertiary Control and Remedial Action Schemes

Tertiary Control. Tertiary control can be used to achieve longer-term, high-level objectives defined by microgrids and the main grid. For instance, reliably and efficiently operating a microgrid in the long term can be ensured by the optimal power flow [32]. The OPF control aims to minimize some objectives such as the microgrid's operational cost, unreliability cost, and environmental cost by optimally dispatching DERs, reactive power resources, transformer tap changer settings, demand response resources, and even droop coefficients. OPF can be performed by a microgrid EMS or in a distributed way that may involve other advanced functionalities such as electricity price forecasting, state estimation, or storage state of charge estimation. Once OPF is solved, the results can be sent to the secondary and primary controllers as set points that can be executed locally.

Microgrid Remedial Action Scheme (microRAS). The remedial action scheme, often referred to as fast load shedding, is used to mitigate highly disastrous events that cannot be covered by primary or secondary controls [33, 34]. Specifically, when a microgrid lacks generation and battery power to restore its frequency to an acceptable range, microRAS will be triggered to shed some of the load. In this situation, if the time required to shed the load is long, the condition will continue to deteriorate and will cause more of the load to be shed, which can lead to system collapse. Thus, it is desirable to have a fast microRAS that sheds the minimum amount of load to maintain system stability. microRAS needs to identify an actual island, calculate the load flow in all relevant parts of the microgrid, and compare the load with the available electrical power inside the island. As soon as a shortage is detected, the load shedding system starts operating and sheds the load that cannot be served. To minimize the amount of load to be shed, the microgrid's EMS can solve a mixed-integer programming (MIP) problem based on real-time information gathered from the microgrid's various components. The amount that needs to be shed will then be communicated to the load. The advantage of microRAS over the traditional frequency-relay-based load shedding is that it responds faster, sheds less load, and leads to minimum outages for electricity users.

2.3   Virtual Synchronous Generator

Among many other microgrid control approaches, the virtual synchronous generator (VSG) is a control scheme that aims to make the inverters emulate the behavior of a synchronous generator as closely as possible [35–37]. The upper part of Figure 2.8 shows the inverter circuits that mimic the energy conversion process of a synchronous machine, where the inductance L_s and the resistance r_s are used to emulate the stator winds of the synchronous machine. The control system for the inverter is designed to produce the reference back electromotive force (EMF) of a synchronous generator such that the following synchronous machine equations can be emulated:

$$\mathbf{v}_{abc} = -r_s \mathbf{i}_{abc} - L_s \frac{d\mathbf{i}_{abc}}{dt} + \mathbf{e}_{abc} \qquad (2.4)$$

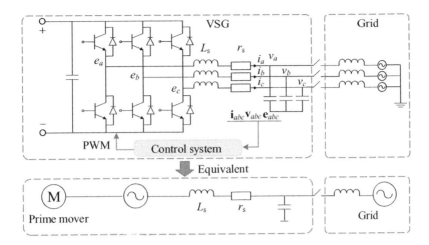

Figure 2.8 Structure of VSG.

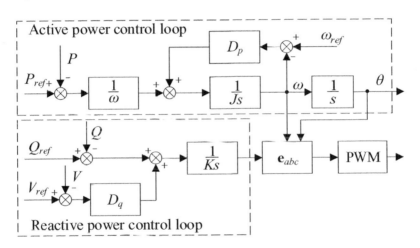

Figure 2.9 Control scheme of VSG.

where $\mathbf{u}_{abc} = [u_a, u_b, u_c]^T$, $\mathbf{i}_{abc} = [i_a, i_b, i_c]^T$ and $\mathbf{e}_{abc} = [e_a, e_b, e_c]^T$ are the generator terminal voltages, currents, and the back EMF, respectively. As shown in Figure 2.9, the VSG control system consists of an active power control loop and a reactive power control loop. The former mimics the behavior of the rotor and speed governor, while the latter mimics the excitation system of a synchronous machine. The back EMF is determined by both the active and reactive power control loops. For a round rotor machine with a constant excitation current, the back EMF can be evaluated by

$$\mathbf{e}_{abc} = M_f i_f \frac{d\theta}{dt} \mathbf{A} \qquad (2.5)$$

where M_f is the mutual inductance between the rotor and the stator, i_f is the rotor excitation current, θ is the rotor angle, and $\mathbf{A} = [sin\theta, sin(\theta - \frac{2\pi}{3}), sin(\theta - \frac{4\pi}{3})]^T$.

The active power control loop produces rotor swing equations including the speed-active power droop effect. The rotor swing equation with frequency-active power droop control can be expressed as

$$\begin{cases} J\frac{d\omega}{dt} = T_m - T_e - D_p(\omega - \omega_{ref}) \\ \frac{d\theta}{dt} = \omega \end{cases} \tag{2.6}$$

where J is the moment of inertia; T_m and T_e are the mechanical and electrical torque, respectively; D_p is the damping factor; ω is the angular speed; and ω_{ref} is its nominal value. By mimicking the rotor swing equations, VSG could provide digitized inertia and damping.

In the reactive power control loop, the voltage-reactive power droop is considered to regulate reactive power according to the deviation between voltage amplitude ΔV and reactive power difference ΔQ. The voltage-drooping coefficient D_q is selected to represent the reactive power change with respect to the voltage changes, i.e. $D_q = -\Delta Q/\Delta V$. Then, this error is added into the tracking error between Q_{ref} and Q to generate $M_f i_f$. Finally, the reference back EMF of VSG can be generated as shown in Figure 2.9.

VSG is still being studied to investigate its capabilities for providing inertia and damping. Various control schemes for VSG have been proposed for potential use cases including the following: damping the DC-side fluctuation in a multiterminal direct current (MTDC) system [38], improving the power transfer capabilities of a weak AC grid [39], and providing VSG functionalities for an interface converter in an islanded hybrid AC/DC microgrid [40]. Recently, a stability analysis has been performed for VSG including the small-signal stability of a power grid with VSG [41], and a parameter optimization has been proposed that aims to improve VSG performance in microgrids [42].

2.4   A Note about DER Modeling

In recent years, several studies have been published that aim to guide readers through the various facets of DER modeling, analysis, and simulation. Hence, instead of covering the detailed modeling of DERs, this chapter will list a few working resources for the modeling of DER units and control:

- General guide to generators and machine modeling:

  - *Power System Stability and Control* [17]
  - *Analysis of Electric Machinery and Drive Systems* [43]

- General guide to DER modeling:

  - *Simulation of Power System with Renewables* [44]

- Photovoltaic power system modeling:

  – *Photovoltaic Power System: Modeling, Design, and Control* [45]

- Wind turbine power system modeling:

  – *Power Conversion and Control of Wind Energy Systems* [46]

- Energy storage system modeling:

  – *Converter-Interfaced Energy Storage Systems: Context, Modelling and Dynamic Analysis* [47]

- Fuel cell modeling:

  – *Modeling and Control of Fuel Cells: Distributed Generation Applications* [48]

References

[1] G. O. Suvire, M. G. Molina, and P. E. Mercado, "Improving the Integration of Wind Power Generation into AC Microgrids Using Flywheel Energy Storage," *IEEE Transactions on Smart Grid*, vol. 3, no. 4, pp. 1945–1954, 2012.

[2] A. Chaouachi, R. M. Kamel, R. Andoulsi, and K. Nagasaka, "Multiobjective Intelligent Energy Management for a Microgrid," *IEEE Transactions on Industrial Electronics*, vol. 60, no. 4, pp. 1688–1699, 2013.

[3] F. R. Yu, P. Zhang, W. Xiao, and P. Choudhury, "Communication Systems for Grid Integration of Renewable Energy Resources," *IEEE Network*, vol. 25, no. 5, pp. 22–29, 2011.

[4] Z. Fan, P. Kulkarni, S. Gormus, C. Efthymiou, G. Kalogridis, M. Sooriyabandara, Z. Zhu, S. Lambotharan, and W. H. Chin, "Smart Grid Communications: Overview of Research Challenges, Solutions, and Standardization Activities," *IEEE Communications Surveys & Tutorials*, vol. 15, no. 1, pp. 21–38, 2013.

[5] J. M. Guerrero, M. Chandorkar, T.-L. Lee, and P. C. Loh, "Advanced Control Architectures for Intelligent Microgrids – Part I: Decentralized and Hierarchical Control," *IEEE Transactions on Industrial Electronics*, vol. 60, no. 4, pp. 1254–1262, 2013.

[6] J. M. Guerrero, L. G. de Vicuña, J. Matas, M. Castilla, and J. Miret, "Output impedance Design of Parallel-Connected UPS Inverters with Wireless Load-Sharing Control," *IEEE Transactions on Industrial Electronics*, vol. 52, no. 4, pp. 1126–1135, 2005.

[7] M. Castilla, L. G. De Vicuña, J. Guerrero, J. Matas, and J. Miret, "Design of Voltage-Mode Hysteretic Controllers for Synchronous Buck Converters Supplying Microprocessor Loads," *IEE Proceedings–Electric Power Applications*, vol. 152, no. 5, pp. 1171–1178, 2005.

[8] H. Han, X. Hou, J. Yang, J. Wu, M. Su, and J. M. Guerrero, "Review of Power Sharing Control Strategies for Islanding Operation of AC Microgrids," *IEEE Transactions on Smart Grid*, vol. 7, no. 1, pp. 200–215, 2016.

[9] J. M. Guerrero, J. Matas, L. G. de Vicuña, M. Castilla, and J. Miret, "Wireless-Control Strategy for Parallel Operation of Distributed-Generation Inverters," *IEEE Transactions on Industrial Electronics*, vol. 53, no. 5, pp. 1461–1470, 2006.

[10] J. M. Guerrero, L. G. de Vicuña, J. Matas, M. Castilla, and J. Miret, "A Wireless Controller to Enhance Dynamic Performance of Parallel Inverters in Distributed Generation Systems," *IEEE Transactions on Power Electronics*, vol. 19, no. 5, pp. 1205–1213, 2004.

[11] J. M. Guerrero, L. Hang, and J. Uceda, "Control of Distributed Uninterruptible Power Supply Systems," *IEEE Transactions on Industrial Electronics*, vol. 55, no. 8, pp. 2845–2859, 2008.

[12] J. M. Guerrero, N. Berbel, J. Matas, J. L. Sosa, and L. G. de Vicuña, "Control of Line-Interactive UPS Connected in Parallel Forming a Microgrid," in *2007 IEEE International Symposium on Industrial Electronics*, pp. 2667–2672, 2007.

[13] M. Castilla, L. G. de Vicuña, J. M. Guerrero, J. Miret, and N. Berbel, "Simple Low-Cost Hysteretic Controller for Single-Phase Synchronous Buck Converters," *IEEE Transactions on Power Electronics*, vol. 22, no. 4, pp. 1232–1241, 2007.

[14] M. Castilla, L. G. de Vicuña, J. M. Guerrero, J. Matas, and J. Miret, "Designing VRM Hysteretic Controllers for Optimal Transient Response," *IEEE Transactions on Industrial Electronics*, vol. 54, no. 3, pp. 1726–1738, 2007.

[15] J. C. Vasquez, R. A. Mastromauro, J. M. Guerrero, and M. Liserre, "Voltage Support Provided by a Droop-Controlled Multifunctional Inverter," *IEEE Transactions on Industrial Electronics*, vol. 56, no. 11, pp. 4510–4519, 2009.

[16] J. M. Guerrero, J. Matas, L. G. de Vicuña, M. Castilla, and J. Miret, "Decentralized Control for Parallel Operation of Distributed Generation Inverters Using Resistive Output Impedance," *IEEE Transactions on Industrial Electronics*, vol. 54, no. 2, pp. 994–1004, 2007.

[17] P. Kundur, *Power System Stability and Control*. McGraw-Hill, 1994.

[18] J. Machowski, J. Bialek, and J. Bumby, *Power System Dynamics: Stability and Control*. John Wiley & Sons, 2011.

[19] J. M. Guerrero, P. C. Loh, T.-L. Lee, and M. Chandorkar, "Advanced Control Architectures for Intelligent Microgrids – Part II: Power Quality, Energy Storage, and AC/DC Microgrids," *IEEE Transactions on Industrial Electronics*, vol. 60, no. 4, pp. 1263–1270, 2013.

[20] J. Schiffer, "Stability and Power Sharing in Microgrids," Ph.D. dissertation, 2015.

[21] A. J. Wood, B. F. Wollenberg, and G. B. Sheblé, *Power Generation, Operation, and Control*. John Wiley & Sons, 2013.

[22] H. W. Dommel, *Notes on Power System Analysis*. University of British Columbia, 1975.

[23] H. D. Vu and J. Agee, "WECC Tutorial on Speed Governors," WECC Control Work Group, 1998.

[24] Y. Li, P. Zhang, L. Ren, and T. Orekan, "A Geršgorin Theory for Robust Microgrid Stability Analysis," in *2016 IEEE Power and Energy Society General Meeting (PESGM)*. IEEE, pp. 1–5, 2016.

[25] A. Engler, "Applicability of Droops in Low Voltage Grids," *International Journal of Distributed Energy Resources*, vol. 1, no. 1, pp. 1–6, 2005.

[26] F. Katiraei, R. Iravani, N. Hatziargyriou, and A. Dimeas, "Microgrids Management," *IEEE Power and Energy Magazine*, vol. 6, no. 3, pp. 54–65, 2008.

[27] C. Wang, J. Wu, J. Ekanayake, and N. Jenkins, *Smart Electricity Distribution Networks*. CRC Press, 2017.

[28] K. De Brabandere, B. Bolsens, J. Van den Keybus, A. Woyte, J. Driesen, R. Belmans, and K. Leuven, "A Voltage and Frequency Droop Control Method for Parallel Inverters,"

in *2004 IEEE 35th Annual Power Electronics Specialists Conference (IEEE Cat. No. 04CH37551)*, vol. 4. IEEE, pp. 2501–2507, 2004.

[29] J. M. Guerrero, J. C. Vasquez, J. Matas, L. G. de Vicuña, and M. Castilla, "Hierarchical Control of Droop-Controlled AC and DC Microgrids – a General Approach toward Standardization," *IEEE Transactions on Industrial Electronics*, vol. 58, no. 1, pp. 158–172, 2011.

[30] A. Bidram and A. Davoudi, "Hierarchical Structure of Microgrids Control System," *IEEE Transactions on Smart Grid*, vol. 3, no. 4, pp. 1963–1976, 2012.

[31] J. W. Simpson-Porco, Q. Shafiee, F. Dörfler, J. C. Vasquez, J. M. Guerrero, and F. Bullo, "Secondary Frequency and Voltage Control of Islanded Microgrids via Distributed Averaging," *IEEE Transactions on Industrial Electronics*, vol. 62, no. 11, pp. 7025–7038, 2015.

[32] Y. Levron, J. M. Guerrero, and Y. Beck, "Optimal Power Flow in Microgrids with Energy Storage," *IEEE Transactions on Power Systems*, vol. 28, no. 3, pp. 3226–3234, 2013.

[33] C. Gouveia, J. Moreira, C. Moreira, and J. P. Lopes, "Coordinating Storage and Demand Response for Microgrid Emergency Operation," *IEEE Transactions on Smart Grid*, vol. 4, no. 4, pp. 1898–1908, 2013.

[34] Y.-Y. Hong, M.-C. Hsiao, Y.-R. Chang, Y.-D. Lee, and H.-C. Huang, "Multiscenario Underfrequency Load Shedding in a Microgrid Consisting of Intermittent Renewables," *IEEE Transactions on Power Delivery*, vol. 28, no. 3, pp. 1610–1617, 2013.

[35] Q.-C. Zhong and G. Weiss, "Synchronverters: Inverters That Mimic Synchronous Generators," *IEEE Transactions on Industrial Electronics*, vol. 58, no. 4, pp. 1259–1267, 2011.

[36] H.-P. Beck and R. Hesse, "Virtual Synchronous Machine," in *2007 9th International Conference on Electrical Power Quality and Utilisation*. IEEE, pp. 1–6, 2007.

[37] K. Visscher and S. W. H. De Haan, "Virtual Synchronous Machines (VSG's) for Frequency Stabilisation in Future Grids with a Significant Share of Decentralized Generation," in *CIRED Seminar 2008: SmartGrids for Distribution*. IET, pp. 1–4, 2008.

[38] C. Li, Y. Li, Y. Cao, H. Zhu, C. Rehtanz, and U. Häger, "Virtual Synchronous Generator Control for Damping DC-Side Resonance of VSC-MTDC System," *IEEE Journal of Emerging and Selected Topics in Power Electronics*, vol. 6, no. 3, pp. 1054–1064, 2018.

[39] A. Asrari, M. Mustafa, M. Ansari, and J. Khazaei, "Impedance Analysis of Virtual Synchronous Generator-Based Vector Controlled Converters for Weak AC Grid Integration," *IEEE Transactions on Sustainable Energy*, vol. 10, no. 3, pp. 1481–1490, 2019.

[40] G. Melath, S. Rangarajan, and V. Agarwal, "A Novel Control Scheme for Enhancing the Transient Performance of an Islanded Hybrid AC-DC Microgrid," *IEEE Transactions on Power Electronics*, vol. 34, no. 10, pp. 9644–9654, 2019.

[41] W. Du, Q. Fu, and H. Wang, "Power System Small-Signal Angular Stability Affected by Virtual Synchronous Generators," *IEEE Transactions on Power Systems*, 2019.

[42] J. Alipoor, Y. Miura, and T. Ise, "Stability Assessment and Optimization Methods for Microgrid with Multiple VSG Units," *IEEE Transactions on Smart Grid*, vol. 9, no. 2, pp. 1462–1471, 2018.

[43] P. C. Krause, O. Wasynczuk, S. D. Sudhoff, and S. Pekarek, *Analysis of Electric Machinery and Drive Systems*. Wiley-IEEE Press, 2002, vol. 2.

[44] L. Kunjumuhammed, S. Kuenzel, and B. Pal, *Simulation of Power System with Renewables*. Academic Press, 2019.

[45] W. Xiao, *Photovoltaic Power System: Modeling, Design, and Control*. John Wiley & Sons, 2017.

[46] B. Wu, Y. Lang, N. Zargari, and S. Kouro, *Power Conversion and Control of Wind Energy Systems*. John Wiley & Sons, 2011, vol. 76.

[47] F. Milano and Á. O. Manjavacas, *Converter-Interfaced Energy Storage Systems: Context, Modelling and Dynamic Analysis*. Cambridge University Press, 2019.

[48] M. H. Nehrir and C. Wang, *Modeling and Control of Fuel Cells: Distributed Generation Applications*. John Wiley & Sons, 2009, vol. 41.

Part II

Networked Microgrids

3 Compositional Networked Microgrid Power Flow

3.1 Challenges of Networked Microgrid Power Flow

Networked microgrids (NMs) unlock microgrids' potential by enabling microgrids to exchange power, by pooling microgrid clusters' generating capacities to support more critical loads, and even by blackstarting the main distribution grid's neighboring energy sources and feeders after a major outage. Though NMs are useful in so many ways, one problem has proven difficult to solve: calculating power flow. Accurate power flow calculations are crucial to planning and operating intentionally islanded NMs, but they can be difficult to obtain.

There are two significant hurdles that make calculating NM power flow particularly difficult. First, the concept of the swing bus no longer applies to an islanded microgrid. It is well known that the existence of a swing bus is the fundamental assumption for traditional power flow calculations. For instance, there must be a swing bus to offer voltage magnitude and angle references in Newton-type power flow algorithms. For distribution load flow methods such as backward/forward sweep, one also needs a root node (e.g., a distribution substation) to serve as a swing bus that has a known voltage angle and magnitude. However, DERs in microgrids are operated by droop control, and the microgrid loads also follow droop characteristics. Second, NMs are subject to frequent changes in configuration and operation modes, especially during the plug-and-play of microgrids or microgrid components. As a result, NMs are often equipped with a secondary control to restore voltages fully and achieve specified power sharing schemes under system changes. Therefore, we shall accurately model the hierarchical control effect to obtain correct NM power flow results.

To bridge the aforementioned gap, we are introducing a novel compositional power flow (ComPF) approach. The salient feature of ComPF is its ability to model the droop characteristics of DERs and loads, to precisely consider power sharing and voltage restoration among networked microgrids, and to handle the plug-and-play of microgrids in a modular fashion.

3.2 Compositional Power Flow

An NM system's power flow is as complex as the system itself. Fortunately, such a system can be better understood and modeled as a "system of systems," i.e., as a

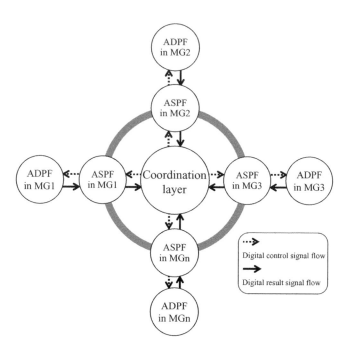

Figure 3.1 Algorithmic framework for ComPF.

composition of subsystems. The concept of composition is key to analyzing composite systems like the ones found in NMs. In the context of NMs, composition can be viewed as the rules that define how microgrids (subsystems) interact. Obviously, composition means those secondary control operations that achieve power sharing and voltage restoration via the tie lines between microgrids. In fact, compositional analysis has been widely used by the formal methods community to perform formal verifications for large dynamic systems [1, 2]. The composition technique is overloaded here to provide a network-level solution for the power flow of multiple microgrids interconnected to achieve some system-level goal. First, we understand the steady-state behavior (power flow) of each individual microgrid by performing an advanced-droop-control-based power flow (ADPF) for individual islanded microgrids. Then, we compose the behaviors of individual microgrids by performing an adaptive-secondary-control-based power flow (ASPF), and eventually the power flow results for the NMs can be obtained. This process is illustrated in Figure 3.1.

3.2.1   ADPF for Individual Islanded Microgrids

When a microgrid is islanded, its DER output may not fully cover its load at the beginning. In that case, the DERs would increase their power generation following the droop control schemes (e.g., P–f/Q–V droop curves) while the loads would decrease following their droop characteristics as illustrated in Figure 3.2. In the following power

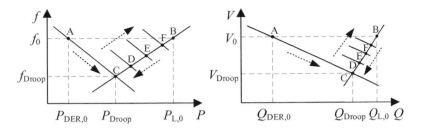

Figure 3.2 Schematics of ADPF and ASPF.

flow formulation, we use the P–F/Q–V droop strategies provided in Figure 3.2 as an example. Formulations with other droop strategies can be derived likewise.

As shown in Figure 3.2, we assume that, in Microgrid i, the DER's initial operation point is A, while the load demand point is B. The purpose of ADPF is to adjust both the DER's generation and load demands according to their droop characteristics as shown in (3.1) and (3.2). By engaging in a step-by-step iteration process, the power generation and loads in Microgrid i will eventually reach an equilibrium at C.

Let the P–f droop coefficient vector be $\mathbf{m}_{P,i}$ and the Q–V droop coefficient vector $\mathbf{m}_{Q,i}$. f^r is the reference frequency that is expected to be restored in Microgrid i. $\mathbf{V_i^r}$ denotes the reference voltages for the DERs in Microgrid i. Assume f and $\mathbf{V_i}$ are the actual frequency and the DER voltages in Microgrid i during a power flow iteration. It should be noted that microgrid frequency f is a global quantity that is the same for the entire microgrid in steady state. $\mathbf{V_i}$, however, are local quantities measured at DER buses in Microgrid i. Note that, if Microgrid i is regulated by droop control, no communication network is particularly necessary. With the preceding discussions, the following droop control strategies in Microgrid i can be implemented in the ADPF process:

$$\Delta \mathbf{P_i} = (f^r - f)/\mathbf{m}_{P,i}, \tag{3.1}$$

$$\Delta \mathbf{Q_i} = (\mathbf{V_i^r} - \mathbf{V_i})/\mathbf{m}_{Q,i}, \tag{3.2}$$

3.2.2   ASPF for Networked Microgrids

At a higher layer than ADPF is ASPF, which is designed to coordinate operations between microgrids. Let us define the powers flowing from Microgrid i to Microgrid j through the tie-lines as $\mathbf{S_{i,j}} = \mathbf{P_{i,j}} + j\mathbf{Q_{i,j}}$. These power interchanges and the loads $\mathbf{S_i}$ in Microgrid i can be combined to calculate a microgrid's voltages. If we assume Microgrid i has a radial structure, then the voltages can be obtained by using the branch current to bus voltage matrix $\mathbf{BCBV_i}$ and the bus injection to branch current matrix $\mathbf{BIBC_i}$ [3], as follows:

$$\mathbf{V_i} = \mathbf{V_i^r} - \mathbf{BCBV_i} \cdot \mathbf{BIBC_i} \cdot \{(\mathbf{S_i} + \mathbf{S_{i,j}})/\mathbf{V_i}\}^*. \tag{3.3}$$

For a microgrid with l branches and n buses or nodes, **BCBV** and **BIBC** can be built using the following computer algorithm [4]:

1: **Create** $(n-1) \times l$ matrix **BCBV** and $l \times (n-1)$ matrix **BIBC**
2: **repeat**
3:      If there is a branch b_m that links buses i and j, then copy the ith column of **BIBC** to the jth column and set $(\mathbf{BIBC})_{mj}$ to 1
4: **until** All branches are screened
5: Obtain **BCBV** by multiplying the transpose of **BIBC** with an impedance matrix **Z** where each main diagonal item is the corresponding branch impedance and all other elements are zeros

---

Power sharing among NMs is the key to the composition of individual microgrids' power flow behaviors. Therefore, it is critical to accurately model and update $S_{i,j}$ in ASPF. The way to update $S_{i,j}$ depends on the NM operating mode. Following are in-depth discussions of two methods for updating $S_{i,j}$, which correspond to two popular operating modes for NMs, i.e., the voltage control mode and the power dispatch mode.

Voltage Control Mode

In the voltage control mode, the voltage of the lead DER bus in Microgrid i is maintained at a specified level. Practical experience shows that, if the leading DER's voltage is treated independently from the interchange of power, then the convergence performance of ComPF can be very poor. This means that, in each iteration, we should adjust power interchanges $S_{i,j}$ based on the voltage deviations at the lead DER bus.

Assume that the active power and the reactive tie-line power flows obtained from the previous iteration are $P_{i,j}^P$ and $Q_{i,j}^P$ respectively. We now compare the reference voltage at the lead DER bus ($V_{DER,i}^r$) with the actual voltage at the same bus ($V_{DER,i}$), bearing in mind that active power is largely determined by the real part of the voltage phasor whereas reactive power depends on the imaginary part. If $real(V_{DER,i}^r - V_{DER,i})$ is negative, then the active power interchange $P_{i,j}$ can be increased to increase the voltage drop in the power delivery process so that the real part of the lead DER's voltage has a better chance of lowering to the desired reference value. Meanwhile, if $imag(V_{DER,i}^r - V_{DER,i})$ is positive, then the reactive power interchange $Q_{i,j}$ should be increased to boost the microgrid's var output so that the reactive part of the lead DER's voltage can be improved, thus enabling it to approach the desired reference value. As the voltage deviations are normalized values per unit, to formally link the power interchange increments with them, we need to scale the voltage deviations by multiplying the power bases and some coefficients so that they have the same units and values comparable to the incremental power interchanges. Here we can use the maximum active and reactive powers of Microgrid i to serve as $P_{base,i}$ and $Q_{base,i}$, respectively, and we can use α_i and β_i as the corresponding coefficients.

Based on the preceding discussions, the voltage control mode associates the interchange of power with the secondary control in the following way:

$$\mathbf{P_{i,j}} = \mathbf{P^p_{i,j}} - \alpha_i \cdot \mathbf{P_{base,i}} \cdot real(\mathbf{V^r_{DER,i}} - \mathbf{V_{DER,i}}), \qquad (3.4)$$

$$\mathbf{Q_{i,j}} = \mathbf{Q^p_{i,j}} + \beta_i \cdot \mathbf{Q_{base,i}} \cdot imag(\mathbf{V^r_{DER,i}} - \mathbf{V_{DER,i}}). \qquad (3.5)$$

Power Dispatch Mode

If NMs operate in the power dispatch mode, those microgrids will be controlled to ensure that the power interchanges across the tie lines meet specified values. Because the power interchange is a function of the voltage across tie lines and the level of tie-line impedance, the voltages at the interface buses need to be adjusted to the proper values so that the specified power interchanges can be achieved. Therefore, the power flow iterations under the power dispatch mode need to emulate the process by which the interface bus voltages are adjusted in order to keep the power interchanges between microgrids constant.

First, power interchanges $\mathbf{S_{i,j}}$ are scheduled and are known throughout the iteration process. If a power flow iteration is obtained by running (3.3), then the voltage changes at interface bus k can be obtained by $\{\mathbf{BCBV_i} \cdot \mathbf{BIBC_i} \cdot \{(\mathbf{S_i} + \mathbf{S_{i,j}})/\mathbf{V_i}\}^*\}_k$. In each microgrid, there is a DER bus serving as the adaptive swing bus. If there is a voltage change at bus k between this iteration and the previous iteration, we need to adjust the reference value of adaptive swing bus $V^r_{i,s}$ and prepare to run another power flow calculation via (3.3) aiming to either reduce the voltage change at bus k or make it zero. It is obvious that the most straightforward way to update the voltage of the adaptive swing bus is to assign the voltage deviation at bus k as the increment for $V^r_{i,s}$. Therefore, the adaptive swing bus voltage in Microgrid i is updated using the following equation:

$$V_{i,s} = V^r_{i,s} + \{\mathbf{BCBV_i} \cdot \mathbf{BIBC_i} \cdot \{(\mathbf{S_i} + \mathbf{S_{i,j}})/\mathbf{V_i}\}^*\}_k. \qquad (3.6)$$

Once the reference voltage of the adaptive swing bus is updated through (3.6), it can be used to calculate the voltages of all the other buses in the ith microgrid by using (3.3). The iteration will continue until the deviations of the interface bus voltages are below a small tolerance such that the scheduled power interchanges are achieved precisely.

By its nature, ASPF is modular, helping protect the privacy of microgrid customers:

- If ComPF is implemented as a class, the control modes can be easily added as function modules in it and can be called and switched between each other upon the user's request.
- ASPF calculations only need a small amount of data to communicate with microgrids. For the voltage control mode, ASPF only needs two things to be sent to the coordination layer so that the power interchanges can be communicated: (1) a sample of DER voltages and power interchanges in the voltage control mode and (2) power interchanges in the power dispatch mode. This could help protect data privacy among microgrids.

3.2.3   ComPF Algorithm

Based on the preceding discussion, the ComPF calculation is divided into a few layers. On the bottom layer is the ADPF calculation for each and every microgrid following the droop characteristics of DERs and loads. On the upper layer is the ASPF layer that updates the interface's power flows. The following *algorithm* summarizes the ComPF iteration process.

1: **Initialize $BCBV_i$, $BIBC_i$, P_i^0, Q_i^0, V_i, V_i^r, f, f^r**
2: **repeat**
3:     Update $P_{i,j}$, $Q_{i,j}$ Eq. (3.4/3.5) or scheduled values
4:     **repeat**
5:         Update ΔP_i, ΔQ_i Eq. (3.1/3.2)
6:         **repeat**
7:             Update V_i Eq. (3.3/3.6)
8:         **until V_i is invariant**
9:         Update V_i, f
10:     **until $V_{i,s}$ and f are invariant**
11:     Update $V_{DER,i}$
12: **until $V_{DER,i}$ or $P_{i,j}$, $Q_{i,j}$ is invariant**

ComPF serves as a general framework for calculating power flow in microgrids equipped with hierarchical controls. If we replace the backward/forward sweep core [5] in the algorithm with the Newton power flow solution [6] or the implicit Z_{bus} Gauss algorithm [7], ComPF can deal with both radial and meshed microgrids [8]. ComPF can be deployed as an essential application in microgrid energy management systems and can also be used to provide accurate initial conditions for the dynamic simulations of NMs.

3.3   Test and Validation of Compositional Power Flow

A 69-bus, three-microgrid NM system (see Figure 3.3) is adopted to validate the effectiveness of ComPF. The base voltage is 12.66 kV, and the base power is selected to be 5,000 kVA [9]. More details of the backbone parameters can be found in Tables 3.1 and 3.2. Four cases are used to evaluate ComPF. *Case A* is designed to validate ADPF, where droop control is used to operate three unconnected, islanded microgrids. Once *Case A* is solved, *Case B* then flips the two switches 3–4 and 9–10 to network those three microgrids. The NM's behaviors are then observed during these natural power interchanges. After *Case B* is solved, both the active and reactive power of Load 61 increase by 50% to create a power imbalance in Microgrid 2. Then, we use *Case C* to verify the capability of ComPF in modeling the voltage control mode to eliminate this power imbalance and increase the efficacy of the power flow results. Finally, *Case D* starts the power flow calculation over again by switching the control

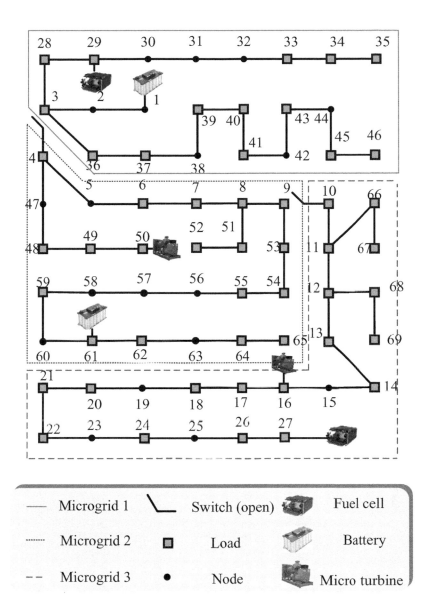

Figure 3.3 A 69-bus test networked microgrid system. DER capacities: 2,160 kVA (bus 1), 1,440 kVA (bus 29), 2,700 kVA (bus 50), 1,800 kVA (bus 61), 2,160 kVA (bus 16), 1,080 kVA (bus 27).

mode to the power dispatch mode to validate the feasibility of ComPF in modeling this type of secondary control.

Case A: Power Flow Calculation for Disconnected Islanded Microgrids: The DERs at buses {1, 29, 50, 61, 16, 27} are controlled using the droop strategy. Their droop coefficients are set as {1.0, 1.5, 0.8, 1.2, 1.0, 2.0}, respectively. Meanwhile, the loads at buses {28, 45, 46, 49, 61, 64, 11, 12} are controlled using the droop strategy as

Table 3.1 Nodal information of test NM system.

Subsystem	Bus number	PL(kW)	QL(kvar)	PG(kW)	QG(kvar)	V_{max}(p.u.)	V_{min}(p.u.)
	1	0.000	0.000	150.0	150.0	1.05	0.95
	2	0.000	0.000	0.000	0.000	1.05	0.95
	3	0.000	0.000	0.000	0.000	1.05	0.95
	28	26.00	18.60	0.000	0.000	1.05	0.95
	29	26.00	18.60	60.00	10.00	1.05	0.95
	30	0.000	0.000	0.000	0.000	1.05	0.95
	31	0.000	0.000	0.000	0.000	1.05	0.95
	32	0.000	0.000	0.000	0.000	1.05	0.95
	33	14.00	10.00	0.000	0.000	1.05	0.95
	34	19.50	14.00	0.000	0.000	1.05	0.95
Microgrid 1	35	6.000	4.000	0.000	0.000	1.05	0.95
	36	26.00	18.55	0.000	0.000	1.05	0.95
	37	26.00	18.55	0.000	0.000	1.05	0.95
	38	0.000	0.000	0.000	0.000	1.05	0.95
	39	24.00	17.00	0.000	0.000	1.05	0.95
	40	24.00	17.00	0.000	0.000	1.05	0.95
	41	1.200	1.000	0.000	0.000	1.05	0.95
	42	0.000	0.000	0.000	0.000	1.05	0.95
	43	6.000	4.300	0.000	0.000	1.05	0.95
	44	0.000	0.000	0.000	0.000	1.05	0.95
	45	39.22	26.30	0.000	0.000	1.05	0.95
	46	39.22	26.30	0.000	0.000	1.05	0.95
	50	384.7	274.5	1400	1000	1.05	0.95
	49	384.7	274.5	0.000	0.000	1.05	0.95
	48	79.00	56.40	0.000	0.000	1.05	0.95
	47	0.000	0.000	0.000	0.000	1.05	0.95
	4	0.000	0.000	0.000	0.000	1.05	0.95
	5	0.000	0.000	0.000	0.000	1.05	0.95
	6	2.600	2.200	0.000	0.000	1.05	0.95
	7	40.40	30.00	0.000	0.000	1.05	0.95
	8	75.00	54.00	0.000	0.000	1.05	0.95
	9	30.00	22.00	0.000	0.000	1.05	0.95
	53	4.350	3.500	0.000	0.000	1.05	0.95
	54	26.40	19.00	0.000	0.000	1.05	0.95
Microgrid 2	50	384.7	274.5	1400	1000	1.05	0.95
	55	24.00	17.20	0.000	0.000	1.05	0.95
	56	0.000	0.000	0.000	0.000	1.05	0.95
	57	0.000	0.000	0.000	0.000	1.05	0.95
	58	0.000	0.000	0.000	0.000	1.05	0.95
	59	100.0	72.00	0.000	0.000	1.05	0.95
	60	0.000	0.000	0.000	0.000	1.05	0.95
	61	1244	888.0	1300	800	1.05	0.95
	62	32.00	23.00	0.000	0.000	1.05	0.95
	63	0.000	0.000	0.000	0.000	1.05	0.95
	64	227.0	162.0	0.000	0.000	1.05	0.95
	65	59.00	42.00	0.000	0.000	1.05	0.95
	51	40.50	28.30	0.000	0.000	1.05	0.95
	52	3.600	2.700	0.000	0.000	1.05	0.95
	27	14.00	10.00	450.0	300.0	1.05	0.95
	26	14.00	10.00	0.000	0.000	1.05	0.95
	25	0.000	0.000	0.000	0.000	1.05	0.95
	24	28.00	20.00	0.000	0.000	1.05	0.95

Table 3.1 (*cont.*)

Subsystem	Bus number	PL(kW)	QL(kvar)	PG(kW)	QG(kvar)	V_{max}(p.u.)	V_{min}(p.u.)
	23	0.000	0.000	0.000	0.000	1.05	0.95
	22	5.000	3.500	0.000	0.000	1.05	0.95
	21	114.0	81.00	0.000	0.000	1.05	0.95
	20	1.000	0.600	0.000	0.000	1.05	0.95
	19	0.000	0.000	0.000	0.000	1.05	0.95
	18	60.00	35.00	0.000	0.000	1.05	0.95
	17	60.00	35.00	0.000	0.000	1.05	0.95
	16	45.50	30.00	300.0	200.0	1.05	0.95
Microgrid 3	15	0.000	0.000	0.000	0.000	1.05	0.95
	14	8.000	5.500	0.000	0.000	1.05	0.95
	13	8.000	5.000	0.000	0.000	1.05	0.95
	12	145.0	104.0	0.000	0.000	1.05	0.95
	11	145.0	104.0	0.000	0.000	1.05	0.95
	10	28.00	19.00	0.000	0.000	1.05	0.95
	68	28.00	20.00	1300	800	1.05	0.95
	69	28.00	20.00	0.000	0.000	1.05	0.95
	66	18.00	13.00	0.000	0.000	1.05	0.95
	67	18.00	13.00	0.000	0.000	1.05	0.95

well with the same droop coefficient 10.0 for both P–F and Q–V schemes. Based on the ADPF results, the frequencies of those islanded microgrids eventually reach 59.5895 Hz, 59.6569 Hz, and 59.8111 Hz, respectively, due to the droop effects. The adjustments of DERs and loads in Microgrid 1 are given as examples in Table 3.3.

Case B: Power Flow Calculation for Freely Networked Microgrids: In this case, we first regulate the frequencies and interface bus voltages in the three microgrids to their nominal values. Then, we connect these three microgrids and observe the synchronization process.

Case C: Operating NMs in Voltage Control Mode: There are two objectives for connecting microgrids using the voltage control mode: (1) to cover the power loads in the system and (2) to adjust the power outputs of particular dispatchable DERs to control their voltages at the required value. Here, we set the terminal voltages of DERs at buses 1, 27, and 50 at 1.01 p.u. The stopping tolerances for the ADPF and ASPF can be set differently. For instance, we may choose to set the ADPF tolerance at 10^{-6} p.u. and the ASDF tolerance at 10^{-3} p.u. If a high accuracy of voltages at interface buses is desired, we can select a smaller stopping tolerance for checking the interface voltages.

Case D: Operating NMs in Power Dispatch Mode: A major reason to network multiple microgrids is to achieve scheduled power interchanges among microgrids so that microgrids with power deficiencies can be supported by those with surplus power. This case is designed to test the accuracy of ComPF when the DER power outputs have to be adjusted to achieve the scheduled power interchanges. In this particular scenario, we assume the DERs in Microgrid 2 have reached their capacity limits so that their output power is restricted to the same level as that in *Case B*. Meanwhile, Microgrids 1 and 3 are regulated to transfer the pre-scheduled exports to meet the increased power demand in Microgrid 2. Microgrids 1 and 3 are assumed to have equal shares in this case.

Table 3.2 Branches information of test NM system.

Subsystem	Bus (from)	Bus (to)	R(Ω)	X(Ω)	Subsystem	Bus (from)	Bus (to)	R(Ω)	X(Ω)
	1	2	0.0005	0.0012		27	26	0.1732	0.0572
	2	3	0.0005	0.0012		26	25	0.3089	0.1021
	3	28	0.0044	0.0108		25	24	0.7488	0.2475
	28	29	0.0640	0.1565		24	23	0.3463	0.1145
	29	30	0.3978	0.1315		23	22	0.1591	0.0526
	30	31	0.0702	0.0232		22	21	0.0140	0.0046
	31	32	0.3510	0.1160		21	20	0.3416	0.1129
	32	33	0.8390	0.2816		20	19	0.2106	0.0690
	33	34	1.7080	0.5646		19	18	0.3276	0.1083
	34	35	1.4740	0.4873		18	17	0.0047	0.0016
Microgrid 1	3	36	0.0044	0.0108	Microgrid 3	17	16	0.3744	0.1238
	36	37	0.0640	0.1565		16	15	0.1966	0.0650
	37	38	0.1053	0.1230		15	14	1.0580	0.3496
	38	39	0.0304	0.0355		14	13	1.0440	0.3450
	39	40	0.0018	0.0021		13	12	1.0300	0.3400
	40	41	0.7283	0.8509		12	11	0.7114	0.2351
	41	42	0.3100	0.3623		11	10	0.1872	0.0619
	42	43	0.0410	0.0478		12	68	0.7394	0.2444
	43	44	0.0092	0.0116		68	69	0.0047	0.0016
	44	45	0.1089	0.1373		11	66	0.2012	0.0611
	45	46	0.0009	0.0012		66	67	0.0047	0.0014
	50	49	0.0822	0.2011		55	56	0.2813	0.1433
	49	48	0.2898	0.7091		56	57	1.5900	0.5337
	48	47	0.0851	0.2083		57	58	0.7837	0.2630
	47	4	0.0034	0.0084		58	59	0.3042	0.1006
	4	5	0.0251	0.0294		59	60	0.3861	0.1172
Microgrid 2	5	6	0.3660	0.1864	Microgrid 2	60	61	0.5075	0.2585
	6	7	0.3810	0.1941		61	62	0.0974	0.0496
	7	8	0.0922	0.0470		62	63	0.1450	0.0738
	8	9	0.0493	0.0251		63	64	0.7105	0.3619
	9	53	0.1740	0.0886		64	65	1.0410	0.5302
	53	54	0.2030	0.1034		8	51	0.0928	0.0473
	54	55	0.2842	0.1447		51	52	0.3319	0.1114
Interface 1–2	3	4	1.7080	0.5337	Interface 2–3	9	10	1.7080	0.5337

Table 3.3 Adjustments of DERs and loads in Microgrid 1 (tolerance: 10^{-6} p.u.).

	Initial values (kVA)	ADPF values (kVA)
DER1	$150.00 + j150.00$	$184.21 + j166.70$
Load28	$26.00 + j18.60$	$22.58 + j16.93$
DER29	$60.00 + j10.00$	$82.81 + j21.21$
Load45	$39.22 + j26.30$	$35.80 + j23.92$
Load46	$39.22 + j26.30$	$35.80 + j23.92$

It should be noted that the voltages are calculated in per unit values, so the voltage differences between nonswing buses and the adaptive swing bus in Microgrid i can be very small between two iterations. To better visualize the convergence process of ComPF, we choose to scale up the voltage differences $\mathbf{r}_i$ by using an $L2$-norm, i.e., $\| \mathbf{c}_i \|_2 = -10/ln(\| \mathbf{r}_i \|_2)$.

Based on the test results, we can obtain several insights:

- By using the ADPF results obtained from *Case A*, one can "reverse engineer" the droop coefficients (see Table 3.3). The inversely calculated droop coefficients are exactly the same as those values preset in the test case. This shows that ADPF can correctly incorporate the droop characteristics in the power flow results.
- The ComPF results for *Case C* shows that the three DER voltages are maintained at their prescheduled values. This means that ComPF can accurately model the effect of the voltage control operations in the NM power flow. A surprise finding is that the voltage quality of Microgrid 2 is somehow not as good as that in *Case B*.

Table 3.4 Droop coefficients calculated via ADPF results for Microgrid 1 (tolerance: 10^{-6} p.u.).

	$\Delta f/\Delta P_i$	$m_{P,i}$	$\Delta V_i/\Delta Q_i$	$m_{Q,i}$
DER1	0.9999	1.0	1.0000	1.0
Load28	10.0027	10.0	10.0185	10.0
DER29	1.4997	1.5	1.5005	1.5
Load45	10.0027	10.0	10.0076	10.0
Load46	10.0027	10.0	10.0084	10.0

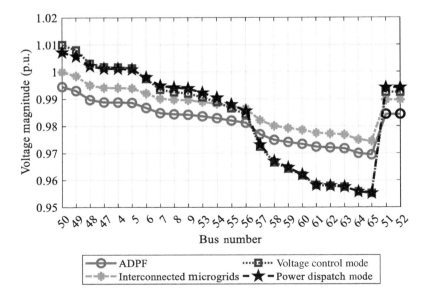

Figure 3.4 Voltage profile of Microgrid 2.

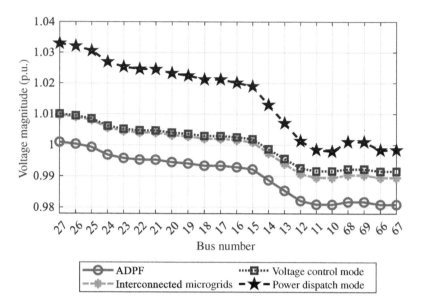

Figure 3.5 Voltage profile of Microgrid 3.

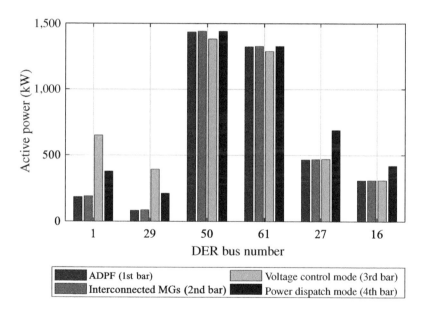

Figure 3.6 Active power generation of DERs.

This is because, even though Microgrid 2's loads are heavy, the voltage control mode cannot increase power imports from its neighboring microgrids to relieve the stressed system voltages. The lack of scheduled power support from neighboring microgrids is illustrated in Figures 3.4 and 3.5.

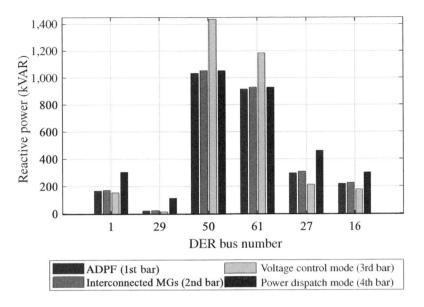

Figure 3.7 Reactive power generation of DERs.

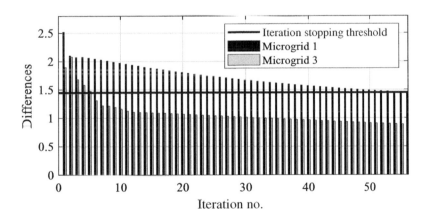

Figure 3.8 Normalized differences of the interface bus voltages during iterations in Case C.

- From the ComPF results for *Case D*, the power shortage in Microgrid 2 is resolved by the properly scheduled power interchanges from the two neighboring microgrids. Thus ComPF can accurately emulate the power dispatch process among networked microgrids. Figures 3.6 and 3.7 show the comparisons of DER outputs in the four cases, *Cases A~D*.
- As an example, Figures 3.8 and 3.9 visualize the iteration processes for *Case C*. Figure 3.8 shows the voltage errors at interface buses, and Figure 3.9 shows the convergence of the two individual microgrids. Both figures show that ComPF has a good performance of convergence.

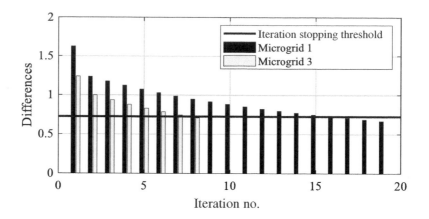

Figure 3.9 Normalized differences of bus voltages for iteration 40 in Figure 3.8.

In summary, ComPF is a privacy-preserving framework for calculating power flows in NMs, which is able to quantify the steady-state effects of primary and secondary controls. It can be used to integrate various microgrid power flow solution algorithms for NMs with arbitrary configurations and electrical characteristics. Therefore, ComPF will be an essential tool in the NM energy management system.

References

[1] S. Tripakis, "Compositionality in the Science of System Design," *Proceedings of the IEEE*, vol. 104, no. 5, pp. 960–972, 2016.

[2] M. Rungger and M. Zamani, "Compositional Construction of Approximate Abstractions of Interconnected Control Systems," *IEEE Transactions on Control of Network Systems*, vol. 5, no. 1, pp. 116–127, 2018.

[3] L. Ren and P. Zhang, "Generalized Microgrid Power Flow," *IEEE Transactions on Smart Grid*, vol. 9, no. 4, pp. 3911–3913, 2018.

[4] J.-H. Teng, "A Direct Approach for Distribution System Load Flow Solutions," *IEEE Transactions on Power Delivery*, vol. 18, no. 3, pp. 882–887, 2003.

[5] G. Díaz, J. Gómez-Aleixandre, and J. Coto, "Direct Backward/Forward Sweep Algorithm for Solving Load Power Flows in AC Droop-Regulated Microgrids," *IEEE Transactions on Smart Grid*, vol. 7, no. 5, pp. 2208–2217, 2016.

[6] P. A. Garcia, J. L. R. Pereira, S. Carneiro, V. M. da Costa, and N. Martins, "Three-Phase Power Flow Calculations Using the Current Injection Method," *IEEE Transactions on Power Systems*, vol. 15, no. 2, pp. 508–514, 2000.

[7] R. D. Zimmerman, "Comprehensive Distribution Power Flow: Modeling, Formulation, Solution Algorithms and Analysis," Ph.D. dissertation, Cornell University New York, 1995.

[8] F. Feng and P. Zhang, "Enhanced Microgrid Power Flow Incorporating Hierarchical Control," *IEEE PES Letters*, Submitted, July 2019.

[9] J. Savier and D. Das, "Impact of Network Reconfiguration on Loss Allocation of Radial Distribution Systems," *IEEE Transactions on Power Delivery*, vol. 22, no. 4, pp. 2473–2480, 2007.

4 Resilient Networked Microgrids through Software-Defined Networking

4.1 Networking Microgrids

Networked microgrids (NMs) consist of a microgrid cluster with cyberphysical interconnections for mutual power and energy interchange. Physically, microgrids in close spatial or electrical proximity can be coupled through utility lines [1], local AC buses [2], or DC links [3]. Once microgrids are successfully networked, longer-term coordination and optimization of NMs at a time scale of minutes, hours, or longer will be possible. For instance, the economic dispatch of NMs can be developed for improving overall microgrid performance through microgrid generation reallocation [4, 5]. If they have the extra reserves to pick up external loads [6] or to black start part of the main grid, NMs can be used to enable the self-healing of distribution networks under power outages. In addition, NMs may also participate in ancillary services such as frequency regulation by injecting desirable real or reactive power into the main grid [7]. In the future, distribution network operators may be able to interact with NMs through integrated NM energy management systems [8, 9]. Today, however, microgrids are often powered by fluctuating and low-inertia distributed energy resources. To achieve the aforementioned envisioned functions, it will be crucially important to enable resilient power sharing within NMs while also ensuring system security. For this purpose, it is essential to equip an ultrafast and reliable cyberinfrastructure for the NM system.

In a single microgrid, either a centralized or a distributed approach can be used to achieve power sharing in tandem with voltage and frequency restoration [10, 11]. In recent years, distributed approaches have started to gain traction because they can largely avoid the so-called single point of failure [12, 13] and help protect customer privacy. Distributed control requiring only local communication has been found to achieve proportional active power sharing and frequency restoration [14]. Similarly, distributed voltage restoration without the consideration of reactive power sharing has also been successfully developed [15]. It should be noted that voltage control and reactive power sharing in DER units may be found to conflict with each other in the context of a droop-based primary control. Among various distributed control methods, the average consensus algorithm has gained popularity for the secondary control in a single microgrid. When used directly for NMs, average consensus requires continuous intensive data transmissions. This may lead to the kinds of bandwidth shortages, congestion, or processor overburdening that could eventually compromise NMs' resilience.

This chapter introduces a software-defined-networking-based (SDN-based) cyber-architecture with a distributed scheme that promises to make NMs more resilient. The unprecedented flexibility and dynamic programmability of SDN [16–18] supports on-the-fly network updates and enables the plug-and-play of microgrids and microgrid components. We will begin this chapter with a short tutorial on software-defined networking before we show how SDN can be used to enhance the resiliency of microgrid networks. This chapter will also introduce event-triggered communication into an SDN-based communication architecture, which will allow a microgrid to share information exclusively with its neighbors when the specific states exceed predefined thresholds. Though event-triggered communication's ability to enable efficient and robust average consensus algorithms for networked control systems has been explored theoretically [19–21], this is the first time it has been integrated into NMs through SDN.

4.2   Software-Defined Networking

Software-defined networking (SDN) has been proposed as means of improving network performance and monitoring because it can both enable programmable network configurations and simplify network management.

4.2.1   Why SDN

Computer networks are notoriously difficult to manage because they involve so many components [22]. Not only are they composed of devices ranging from switches to routers and middleboxes (e.g., firewalls), the software running on those devices is complicated, protected, and provided by different vendors. These complexities mean that users have little latitude for customizing their networks' functionalities. Though the standardization and deployment of new network protocols promises to ease these issues, this will take time. What's worse, individual network devices are managed through configuration interfaces which also vary between different vendors, making it very difficult to manage the network. All of these factors have slowed innovation, increased network complexity, and elevated the costs associated with running a network. This chapter argues that SDN can address these issues by opening up more room for innovation and enabling networks to be managed more efficiently.

4.2.2   SDN Architecture

SDN's most important innovation is its separation of the data plane from the control plane [23]. The network's switches are connected to the SDN controller, which can then be programmed to provide flexible functions to support various QoS requirements. Because the network's switches are capable of detecting network congestion

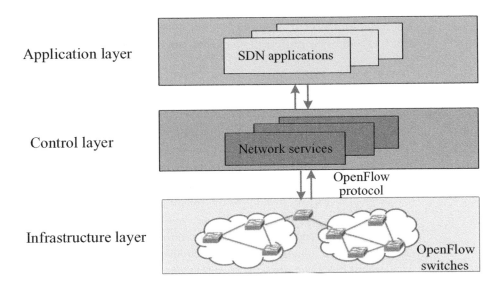

Figure 4.1 SDN architecture.

and failures, they can provide the SDN controller with the information it needs to react to those events in real time.

SDN architecture generally includes three layers, as shown in Figure 4.1.

The **infrastructure layer** is composed of various networking devices that are used for forwarding and processing data for the network. These networking devices receive commands from and execute the rules circulated by the upper-layer SDN controller.

The **control layer** is a control plane that contains the SDN controller. An SDN controller is a framework that receives both instructions and requests from the SDN application layer (upper layer) and relays them to the network's devices. The SDN controller also collects information about the network from the networking devices and tells the SDN application layer about statistics and events occurring in the infrastructure layer. The SDN controller is exposed to two types of interfaces – Northbound and Southbound interfaces. The Southbound interface refers to the connection between the SDN controller and the physical networking devices, while the Northbound interface refers to the connection between the controller and the applications. In short, these interfaces define the communication between the applications, the controllers, and the networking infrastructure.

The **application layer** is the layer where innovative applications can be developed by leveraging all the network's information, including its topology, state, and statistics. These applications may include network management or analytics applications for various systems. For example, in networked microgrids, an event-triggered application can be built to monitor events in the microgrid and facilitate power sharing between networked microgrids.

Unlike traditional networks that are merely application-aware, this SDN architecture gives the applications more information about the state of the network from the controller. Moreover, it uses open protocols for SDN controllers and network switches, which expedites the process of innovating and developing new applications.

4.2.3   OpenFlow

As mentioned earlier, the SDN controller controls the network's devices via its Southbound interface. OpenFlow is a prominent example of the Southbound interface API [24]. With the OpenFlow protocol, the controller can add, update, and delete flow entries in the flow tables of an OpenFlow switch either passively (in response to packets) or proactively. An OpenFlow switch can have one or multiple flow tables that define the rules for dealing with different packets (e.g., forwarding, dropping, or modifying).

The flow table contains different flow entries. As shown in Figure 4.2, the flow entry defines the matching rules and actions for managing data traffic. It consists of matching fields, actions, statistics, flow entry priorities, and timeout settings. The matching field defines the rules to match against the packet header. For example, it may contain the ingress switch port, the packet source, and the destination addresses. The actions component establishes the instructions for dealing with the packet in a way that matches all the values in the matching fields. The actions component includes sending the packet to a certain port or group of ports, dropping the packet, and sending it to the SDN controller to decide. The statistics part will count the total number/bytes of matching packets. Each flow entry needs to be assigned according to its priority (ranging from 0 to 65,536), which defines the matching search sequence in the flow

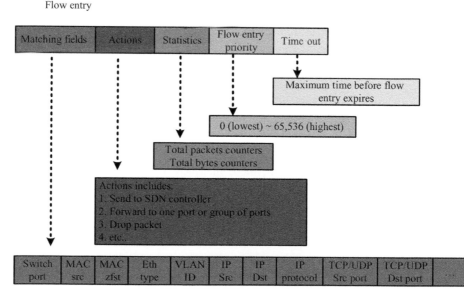

Figure 4.2 OpenFlow flow entry.

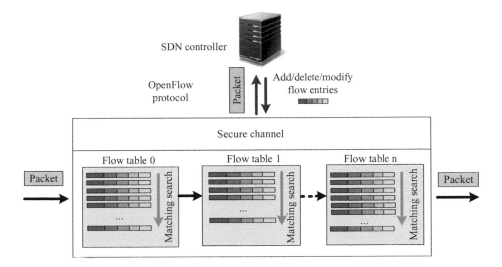

Figure 4.3 SDN pipelines.

table. The higher the number, the higher the priority the flow entry has during the matching search. The lowest priority 0 is assigned to the last entry that needs to be checked. Finally, the timeout field defines the maximum length of time or idle time before the flow entry expires and is discarded.

If an OpenFlow switch has more than one flow table, they are organized in the form of a pipeline, where the packet is allowed to go through different flow tables. In other words, the pipeline processing determines how packets and those flow tables would interact (see Figure 4.3).

When a packet enters an OpenFlow switch, it searches for the flow entry in the first flow table. If the packet can be matched with a flow entry, it will be processed based on the sets of instructions defined in the flow entry. If there is a table forward instruction among the sets of instructions, the packets will be forwarded to a further flow table. Pipeline processing is unidirectional, meaning it only moves forward rather than going backward. For example, a flow entry can only forward a packet to a flow table whose table number is greater than its own. If the instructions set does not define forwarding the packet to another flow table, the pipeline processing will end. When it stops, packets are processed and forwarded based on their associated set of operations. If there is no flow entry that can match the packet, this is called a table-miss situation. The manner in which a table-miss packet is processed relies on the flow table settings. The table-miss flow entries in the flow table can define how to process those packets: it may drop them or forward them with packet-in messages to let the SDN controller process them. The SDN controller can be programmed to handle those packets and to add, delete, or modify flow entries in the OpenFlow switch in the event that this type of packet is received in the future.

4.2.4   SDN-Based Microgrid Communication Architecture

The SDN-based communication architecture for microgrids has been proposed in [25]. The flexible and programmable SDN can be utilized to simplify the management of microgrid communication. As this architecture abstracts the network infrastructure from the application layer applications (for instance, various coordination and control functionalities), it can also simplify the process of developing applications for microgrids.

Figure 4.4 illustrates the SDN-based microgrid communication architecture. It contains four layers: the microgrid layer, the infrastructure layer, the control layer, and the application layer. The microgrid layer includes one or more microgrids. The infrastructure layer contains a group of SDN-capable switches and the links between them. The control layer provides a logically centralized control of the network through the SDN controller. The application layer contains the various applications that run for microgrid control and optimization, such as the primary and secondary control,

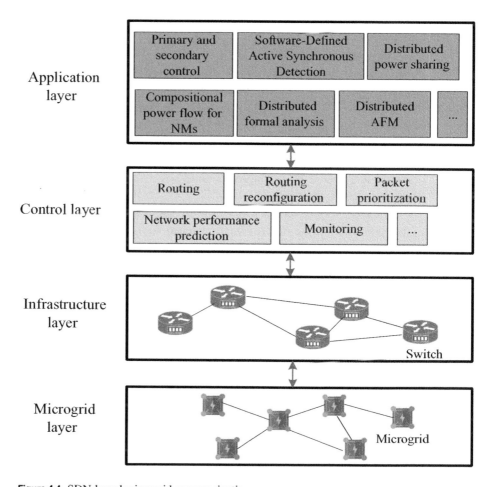

Figure 4.4 SDN-based microgrid communication.

Software-Defined Active Synchronous Detection (SDASD), distributed power sharing, compositional power flow for networked microgrids, distributed formal analysis, and distributed active fault management (AFM). In the preceding architecture, many functionalities can be embedded in the control layer. The functionalities that are most relevant in meeting the communication requirements of microgrids are as follows:

Routing. In the SDN-based architecture, routing is determined by the SDN controller in the control layer and is based on the global network information provided by a monitoring service (to be described later). For a particular flow, the optimal route can be computed proactively (i.e., before receiving any packet) or passively (i.e., after receiving a packet) by solving an optimization problem that is formulated based on the source, the destination, the graph of the network, and the QoS requirement of the flow. For instance, for a flow of small control packets with stringent delay requirements, the optimal route is the one with the shortest delay from the source to the destination; for a flow of large data packets that require a high throughput with a relaxed delay requirement, the optimal route is the one that provides the highest bandwidth. Once the route is determined, each switch on the route simply looks up the forwarding rule (based on the header of the packet) to determine which outgoing port should be forwarded a packet.

Route Reconfiguration. The SDN controller will recalculate the route of a flow when it detects or predicts major changes in the network (informed by the monitoring service). For instance, when the performance of a network link degrades significantly, the SDN controller will recalculate the routes of the flows that traverse this link, and it will push new forwarding rules to relevant switches so that the flows will be rerouted to another path. This can lead to much faster path switching than would be possible under distributed convergence-based routing protocols (e.g., Routing Information Protocol [RIP]).

Packet Prioritization. In addition to routing, queuing mechanisms can be leveraged in SDN switches to provide additional QoS support. SDN switches from some vendors (e.g., HP OpenFlow switches [26]) support multiple strict-priority queues for each physical port. Source/destination port numbers or IP ToS (Type of Service in the Internet Protocol header) fields can be used to differentiate flows with different priorities. To prioritize certain types of flows (e.g., the flows used to transfer microgrid control messages), the SDN controller will push forwarding rules to the switches to map these types of flows to high priority queues.

Monitoring. The monitoring service provides global information about the network to higher-level control programs (e.g., routing and route reconfiguration). This service needs to continuously monitor the network with low overhead. Standard techniques for network monitoring include active probing approaches and passive "capture-and-analyze" approaches (e.g., SPAN and NetFlow). Monitoring techniques for software-defined networking [27–30] include both active and passive probing. Each of these has its advantages and disadvantages: active probing approaches do not require additional instrumentation, but they do inject measurement traffic into the network, which can lead to high overheads. Meanwhile, passive "capture-and-analyze" approaches do not introduce additional traffic but may need additional

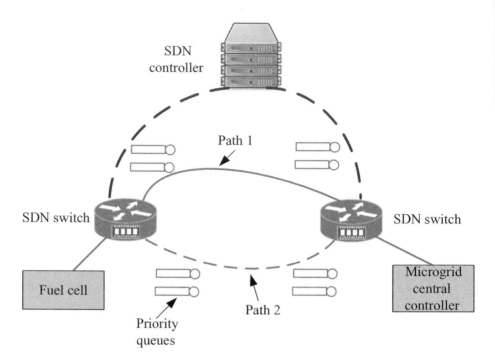

Figure 4.5 An example to illustrate packet prioritization and multipath routing.

instrumentation. Finally, even though SDN-based monitoring techniques involve lower overheads and require no additional instrumentation, they have limited capabilities. In practice, a combination of the approaches described in this section may be needed depending on the requirements of the physical system.

Figure 4.5 provides an example [25] that illustrates the functionalities discussed previously. Consider two nodes – a fuel cell generator and microgrid central controller – in a microgrid. Each node is connected to an SDN switch. Suppose that there are two network paths between these two switches. For the sake of simplicity, let us suppose that there are two flows between this pair of nodes: a control flow that contains small periodic control packets that are sent in short intervals (e.g., every 50 ms, and the delay requirement is in milliseconds) and a data flow that contains large data packets that are sent following a Poisson distribution where the delay requirements is a few seconds. Initially, the SDN controller calculates that the optimal route for both the control and the data flows is the first path, with the control flow being placed in the top priority queue and the data flow being placed in the second priority queue. At a later time, suppose the delay on the first path becomes too long for the control flow. Then the control flow will be rerouted to the second path.

Using SDN for microgrid communication is not straightforward, which means that each microgrid's specific characteristics and requirements need to be considered. Traffic characteristics in microgrids differ significantly based on whether the microgrid in question is part of a home network [31, 32], a data center [33, 34], or a university/enterprise network [35, 36]. In addition, the delay requirements for certain types of flows in microgrids are much more stringent than those scenarios. This communication architecture provides an interface to develop a set of customized techniques to meet the needs of microgrid communication. Specifically, in routing, the controller needs to determine what types of flows need proactive routing and what types of flows can be routed reactively. In route reconfiguration, the range of latencies for switching from one network path to another is needed for different networking technologies and settings, and the SDN controller needs to know how to maintain consistent route updates on multiple switches. In monitoring, it is necessary to quantify the accuracy and overhead of a set of monitoring techniques to determine the best combination of monitoring techniques. Performance prediction techniques (e.g., based on time-series analysis, change-point detection, machine learning) that can predict the onset of performance problems (e.g., congestion) will also be used to proactively reconfigure routes.

4.3   Distributed Power Sharing for Networked Microgrids

We assume that local controllers (LCs) are equipped with inverter-interfaced DERs in each microgrid. When NMs are islanded, droop controls implemented in LCs can automatically adjust DER power outputs to reduce the voltage and frequency variations caused by generation-load imbalances. To achieve power sharing within an individual microgrid (local power sharing), an effective approach is to apply a secondary control such as the distributed averaging proportional-integral (DAPI) control [14] to microgrid LCs, aiming to better restore voltages and frequency. Here we first recapitulate droop control and a local power-sharing algorithm for a single microgrid, and then discuss global power sharing among microgrids through an average consensus approach. This two-layered power-sharing scheme has its limitations, which motivated our design for the event-triggered communication scheme described in Section 4.4.

4.3.1   Droop Control and DAPI Control

Consider an NM system formed by a total of N microgrids labeled as $\mathcal{V} = \{1, \ldots, N\}$. For the ith microgrid, there are N_i dispatchable DERs (e.g., diesel generators, combined heat and power units, microturbines, batteries, etc.) and programmable loads, indexed by $\mathcal{V}_i = \{1, \ldots, N_i\}$. We consider a decentralized scenario where the communication network is undirected. The cyberinfrastructure for the ith microgrid can be described as a graph $\mathcal{G}_i = \{\mathcal{V}_i, \mathcal{E}_i, \mathcal{A}_i\}$, where $\mathcal{E}_i \subseteq \mathcal{V}_i \times \mathcal{V}_i$ is the edge set referring to the cyberconnections between DERs in microgrid i, and $\mathcal{A}_i$ is a binary adjacent matrix with elements $\{a_{mn}^i\}$. a_{mn}^i is 1 if the edge $\{m,n | m,n \in \mathcal{V}_i\}$ exists; otherwise, it is 0. Thus, the cybernetwork for NMs can be represented as $\mathcal{G} = \{\mathcal{V}, \mathcal{E}, \mathcal{A}\}$. The communication between microgrids can then be expressed as $\mathcal{E} \subseteq \mathcal{V} \times \mathcal{V}$ while $\mathcal{A}$ is the adjacent matrix for the NMs.

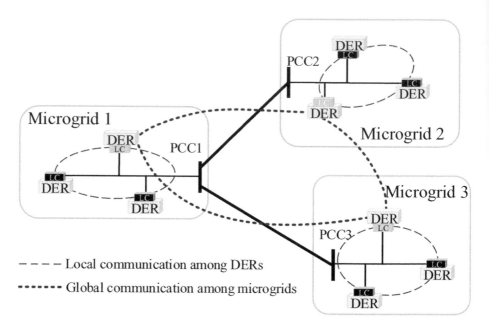

Figure 4.6 A schematic NM system with three microgrids. (LCs in gray: LCs of the leader DERs; LCs in black: LCs of the follower DERs.)

Each individual microgrid already has a local connected communication network. To set up communications with other microgrids or external entities, a dedicated interface is needed. This can normally be achieved by establishing communications among transceivers of LCs. Usually, one of the DERs serves as the leader DER for the ith microgrid, and its transceiver is designated as an interface to allow the microgrid to share information with other microgrids. The leader DER should be able to quickly respond to the request from other microgrids for power support. Therefore, practically, we can select the DER with the largest capacity as the leader DER. Without a loss of generality, the leader DER can be designated as the first DER in the node set $\mathcal{V}_i$. The DER set for the ith microgrids can thus be expressed as $\mathcal{V}_i = \{\mathbf{1}, 2, \ldots, N_i\}$, where the bold $\mathbf{1}$ refers to the leader DER. As an example, an islanded three-microgrid NM system is shown in Figure 4.6.

A widely used droop control expression for the jth DER in the ith microgrid is as follows:

$$f_j = f^* - m_j(P_j - P_j^*) = f_j^* - m_j \Delta P_j, \quad j \in \mathcal{V}_i \tag{4.1a}$$

$$U_j = U_j^* - n_j(Q_j - Q_j^*) = U_j^* - n_j \Delta Q_j, \quad j \in \mathcal{V}_i, \tag{4.1b}$$

where f_j and U_j are the frequency and voltage magnitude of the jth DER, f^* and U_j^* being their references. P_j and Q_j indicate active and reactive power with P_j^* and Q_j^* being their references and ΔP_j and ΔQ_j as the corresponding power error inputs for the droop controller. m_j and n_j are the $P - f$ droop coefficient and the $Q - U$ droop coefficient, respectively. Oftentimes the $P - f$ droop coefficients for DERs can be

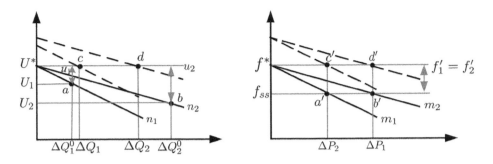

Figure 4.7 Steady-state DAPI performance shown on droop curves.

set to be inversely proportional to their power ratings to achieve proportional power sharing among those DERs.

Frequency synchronization is guaranteed if the active power flows are less than the physical maximums, which is proved by leveraging tools from the theory of coupled oscillators [37, 38]. For the voltage droop control, a major issue is that the steady-state operation point is significantly affected by the line impedance between the inverter-controlled DER and the point of common coupling (PCC). Existing advanced reactive power control methods such as the error compensation control [39] and the adaptive impedance control [40] are mostly centralized solutions. Therefore, this chapter focuses on the accurate sharing of active – rather than the reactive – power, allowing a trade-off between voltage regulation and reactive power sharing in DAPI.

It is well known that droop control will still leave voltage magnitudes and frequency with deviations from their set points after responding to load changes. Figure 4.7 illustrates the $Q - U$ and $P - f$ droop characteristics with equilibrium points a, b, a', and b'. This means that if droop curves are programmed in the DER inverters, in the steady state, the frequency and voltage would deviate from the nominal values in response to the load variations. For the $P - f$ droop, a' and b' can be calculated through (4.1a), resulting in a steady-state synchronized frequency $f_{ss} = f^* - \frac{\sum \Delta P_j}{\sum 1/m_j}$. However, as previously mentioned, for the $Q - U$ droop, it is nontrivial to obtain operation points a and b because they are a function of the droop coefficient, the load level, and the line impedance. To eliminate the deviations in voltage and frequency, we need to introduce a secondary control.

Here we introduce a popular distributed secondary control, that is, the DAPI controller, which is a proportional integral controller based on the average consensus approach. This DAPI controller can effectively mitigate deviations in frequency and voltages which cannot be fully addressed by droop control. Assume the secondary frequency control signal and the secondary voltage control signal are f'_j and U'_j, respectively. For the the jth DER, its neighboring DERs form a set $\mathcal{N}_j \subseteq \mathcal{V}_i$. The rule of thumb to figure out the neighboring DER set of a particular DER is to determine those DERs that have communication links with them. Mathematically, there is an adjacent matrix A_i that corresponds to the local communication network of a microgrid i. Therefore, $\mathcal{N}_j$ can be built up by searching the nonzero elements on the jth row or column of A_i. Within the secondary frequency and voltage control signals, there are two parameters α_j and γ_j, which are associated with frequency and voltage restoration, respectively. There are also two control parameters β_{lj} and δ_{lj} for active and reactive power sharing, respectively. Tuning the parameters to achieve ideal performance, however, can be nontrivial [14]. As can be seen in Figure 4.7, the effect of the DAPI controller in steady-state is that operation points a, b, a' and b' obtained with droop control would be lifted to c, d, c', and d', respectively. It shows that f' of all DERs can reach a unified value whereas U' may be unequal between DERs due to the impedance mismatch (Figure 4.7a illustrates just one possible result). Mathematically, the DAPI for the jth DER in the ith microgrid can be formulated as follows:

$$f_j = f^* - m_j(P_j - P_j^*) + f_j', \quad j \in \mathcal{V}_i \tag{4.2a}$$

$$f_j' = -\int \left\{ \alpha_j(f_j - f_j^*) + \sum_{l \in \mathcal{N}_j} \beta_{lj}(f_j' - f_l') \right\}, \quad j \in \mathcal{V}_i \tag{4.2b}$$

$$U_j = U_j^* - n_j(Q_j - Q_j^*) + U_j', \quad j \in \mathcal{V}_i \tag{4.3a}$$

$$U_j' = -\int \left\{ \gamma_j(U_j - U_j^*) + \sum_{l \in \mathcal{N}_j} \delta_{lj}(Q_j/Q_j^* - Q_l/Q_l^*) \right\}, \quad j \in \mathcal{V}_i. \tag{4.3b}$$

4.3.2   The Global Layer of Active Power Sharing for Networked Microgrids

By using DAPI, any single microgrid in an NM community will be capable of distributed power sharing without relying on a central controller or the one-to-all communication. On a higher level, however, information shall still be shared among microgrids to allow a part of or all microgrids participating in NM-level power exchanges to achieve the expected benefits of NMs. To minimize the cost of networking microgrids, the best practice is to devise a global cyber layer that can support the plug-and-play of microgrids. This global layer shall ensure that local controllers and their parameters within any single microgrid remain unaltered, which not only retains the best local performance of individual microgrids after they are physically networked but also minimizes the capital and operational expenses for interconnecting those microgrids.

Over the years, the author and his team have been developing a software-defined architecture to ensure microgrids can be readily networked at both the physical layer and the control layer in a simple and economically effective way, transforming traditionally isolated local microgrids into interconnected smart microgrids capable of being reconfigured rapidly to achieve the desired resiliency, elasticity, and efficiency. In what follows, we will use the software-defined networks philosophy to develop an innovative way to achieve fast global power sharing. To gain better insights without losing generality, a few assumptions are made in the following discussions:

- The microgrids are coupled via AC lines.
- DERs within the same microgrid are more strongly coupled than those locating in separate microgrids.
- The microgrids do not share reactive power.

It is reasonable to make the first assumption because using existing distribution utilities facilities, such as medium- or low-voltage feeders, is generally more affordable than installing new facilities such as inverter-based DC links for connecting multiple microgrids. As such, most existing research assumes that medium-voltage distribution feeders are being used as the backbone for NMs [41, 42]. The second assumption is generally validated, as DERs in a microgrid are almost always linked with external microgrids through the use of extra feeders. As discussed earlier, proportional active power sharing can be achieved by the local droop controller via frequency synchronization. The time constant of this synchronization depends partly on the strength of the electrical links between any two DERs. Thus, the closely coupled DERs normally would converge faster than the loosely connected ones. The reason why reactive power sharing is not discussed in this book is twofold: First, based on the small signal analysis of DAPI-based reactive power sharing [14], the dissimilarity among DERs may cause instability. Second, it is electrically unfavorable to transfer reactive power to neighboring microgrids because severe voltage issues may occur. Instead, it is better to provide local reactive power compensations such as a shunt capacitor or a Distribution Static Synchronous Compensator (D-STATCOM). For these reasons, global reactive power sharing is disregarded in the current scope of research. In the future, if reactive power can be traded as a commodity among NMs and the reactive power pricing mechanism is established, reactive power sharing among NMs might become a worthwhile topic.

A direct way to achieve global active power sharing is to implement it through leader DERs. For the ith microgrid with its neighboring microgrids denoted by $\mathcal{K}_i \subseteq \mathcal{V}$, the global active power sharing mechanism is modeled by

$$f_1' = -\int \left\{ \alpha_1(f_1 - f^*) + \sum_{l \in \mathcal{N}_1} \beta_{l1}(f_1' - f_l') \right\} - \int \left\{ \sum_{k \in \mathcal{K}_i} \eta_{ki}(f_1' - f_k') \right\}, 1 \in \mathcal{V}_i.$$

(4.4)

In comparison to (4.2b), the difference is that there is an additional term for global power sharing, that is, dispatching power among neighboring microgrids. Here η_{ki} is the parameter for such an error item associated with global active power sharing. The local controls for the follower DERs remain unchanged. The two-layered structure is illustrated in Figure 4.8.

Two-layered power sharing, in steady state, can equalize the secondary frequency variable f' for DERs within a local microgrid and the neighboring microgrids with which it is exchanging power flows. As shown in Figure 4.7b, in a steady state, the microgrid stabilizes at $f_j = f^*$ and $f_j' = f_l', l \in \mathcal{N}_j$. With the global control, DERs' power outputs are further regulated to achieve $f_j' = f_k', k \in \mathcal{K}_i$ and thus $f_j' = f_k' = f_l', l \in \mathcal{N}_j, k \in \mathcal{K}_i$. Since each DAPI on the local layer corresponds to a

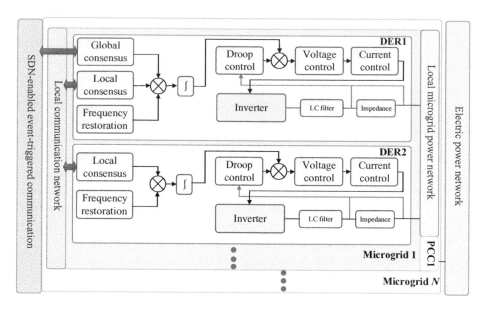

Figure 4.8 Two-layered power sharing scheme for NMs.

communication subgraph ($\mathscr{G}_i, i \in \mathscr{V}$) that is connected, the preceding process applies to DERs in any microgrid i and its neighboring microgrids. This is the process by which power sharing among NMs is achieved.

Two-layer power sharing for NMs is a generalization of DAPI-based power sharing for a single microgrid. Power sharing involving some or all microgrid inverters is guaranteed stable for a droop-control-regulated microgrid [13]. Intuitively, the same conclusion should apply to NMs. However, the AC lines between some of the micro-grid's neighbors may introduce large impedance that physically hinders power-sharing among those microgrids and deteriorates the convergence of the power-sharing process for NMs. Rather than involving all of its neighbors in the global power-sharing layer, it is desirable for each microgrid to identify a subset of neighboring microgrids such that fast power exchange is made possible. Searching for the proper neighboring microgrid set for each microgrid, therefore, becomes a necessary preprocessing technique to enable scalable power sharing and the plug-and-play of NMs.

In addition, the level of communication needed to facilitate global power sharing for a growing NM system requires that additional investments be made in extra bandwidth and upgrades. NMs' communication network also suffers from possible congestion, delays, packet losses, link failures, or attacks that can compromise NMs' performance or even cause system collapse. As introduced in Section 4.2, SDN provides potent solutions for these challenges. Moreover, the programmability of SDN provides an opportunity to enable plug-and-play in NMs. Therefore, a software-defined, flexible communication infrastructure adaptive to the dynamic cyberphysical configuration of NMs will be pursued to enable resilient power sharing and to mitigate the risks and reduce the costs of global communication.

4.4   SDN-Enabled Event-Triggered Communication

Our technical solutions involve designing an SDN-based communication infrastructure and then establishing an event-triggered scheme for the power sharing on the global layer. The performance is further enhanced by sharing power only among selected microgrids that are within short electrical distances. It is known that high latency in microgrids communication will severely affect centralized secondary control performance [43, 44]. The average consensus algorithm may suffer from divergence if the latency is above a certain threshold [45]. The author's team has developed the SDN technology necessary to enable a fast failover, traffic prioritization, and network delay guarantee for guaranteed microgrid communication performance [25]. In addition to this, we are also introducing an electrical distance approach for selecting neighboring microgrids. This forms the basis of the event-triggered microgrid control that is empowered by the SDN architecture.

4.4.1   Sharing Power with the Nearest Neighbors

Each microgrid within an NM system is interfaced with an external power grid through a PCC as shown in Figure 4.8. The strength of the microgrid coupling can be measured by electrical distance, a concept used by transmission system engineers for system partitions [46]. Through $P - f$ droop, active power changes due to frequency error and subsequent phase angle differences. If each microgrid holds desirable voltages through the local secondary control without needing to supply reactive power to neighbors, similar to the traditional DC load flow, the susceptance matrix could somehow represent the sensitivity of active power outputs with respect to voltage angles. Therefore, one can use reactance to measure electrical distances among microgrids. Assume $\mathbf{B}$ is the susceptance matrix of the NM backbone grid that links microgrids and let its pseudo inverse be $\mathbf{B}^+$. Thus the electrical distance d_{mn} between PCC buses m and n may be estimated via (4.5) [46]:

$$d_{mn} = (\mathbf{B}^+)_{mm} - (\mathbf{B}^+)_{mn} - (\mathbf{B}^+)_{nm} + (\mathbf{B}^+)_{nn}. \tag{4.5}$$

The electrical distance matrix $D = \{d_{mn}\}$ can subsequently be formed. By searching the K smallest nondiagonal elements in the ith row, one can determine the so-called K-nearest neighbors for microgrid i. This means that, if a microgrid requests power inputs from the NM system, the global active power sharing control will be directly applied to the K-nearest neighbors, which is a $K-$microgrid cluster centering around that microgrid with the power deficiency.

4.4.2   Event-Triggered Communication and Control through SDN

Obviously, power sharing will be requested only if the microgrid load goes beyond a threshold, creating an emergency state in a microgrid. The threshold can be set using various criteria, such as the available resources of NMs (e.g., the capacity reserve of each individual microgrid) or based on microgrid demand. For instance, one may use

20% of the nominal capacity of the DERs in a microgrid as a threshold to trigger a power-sharing request.

Event-Triggering Logic

Establishing an automatic event-triggering mechanism is key to event-triggered communication. Two basic types of events, E_1 and E_2, are introduced for NM power sharing, where E_1 denotes a global power-sharing request initiated from local microgrids, and E_2 is the request for event clearance after power sharing is settled. Local microgrid controllers are responsible for detecting these events, which are then sent to the SDN controller.

Two conditions may trigger E_1: a significant increase in local load or its recovery. To establish an E_1 signal, the active power error signal $\Delta P_j = P_j - P_j^*, j \in \mathcal{V}_i$ from the droop controller of the jth DER is compared with the predefined threshold P_{th}^i, which has been broadcast to all DERs in the particular microgrid. If $\Delta P_j > P_{th}^i$, this means that a sudden load surge beyond the predefined threshold has occurred, which will lead to a power deficiency in a microgrid. As a result, the local controller will generate an E_1 signal to initiate a request for global power sharing. Otherwise, if $\Delta P_j < -P_{th}^i$ is detected after the load increase has occurred, this means the load has been recovered and an E_1 request still needs to be sent so that the global power-sharing process can be initiated to reduce the import of power from neighboring microgrids.

E_2 is triggered when power sharing is completed and the power deficiency emergency has been addressed. To establish an E_2 signal, the frequency error signal $\Delta f_j = f_j - f_j^*, j \in \mathcal{V}_i$ is used in that frequency restoration is a result of the power-sharing process. When the absolute value of Δf_j is recovered to $\Delta f_j^{th} = m_j * (P_{th}^i)/(K+1)$, the load variation is relieved to a tolerable level. Then the SDN controller can cancel out the global communication established. Note that the detection sequence is important and has to be taken care of correctly. Specifically, an E_1 event should be detected only after the power deficiency occurs, while an E_2 event should only be triggered after an E_1 event exists. To maintain the correct logic, three flags are set in the controller, as illustrated in Figure 4.9. Once triggered, the $E1$ or $E2$ requests are to be processed by the SDN controller as discussed in the following.

SDN-Enabled Event-Triggered Communication and Control

The SDN controller manages the microgrid communication network in a centralized way, which provides it with full access to and programmability on all SDN switches. The SDN controller stores and maintains three tables, including the IP address table ($\mathbf{T}_1$, shown in Figure 4.10), the K-nearest neighbor table ($\mathbf{T}_2$), and the communication connectivity table ($\mathbf{T}_3$). Once there are changes in the NMs, structure (for instance, the plug-in of a microgrid) or in the microgrid communication network (e.g., the adding, changing, or removal of an IP address), those tables are updated correspondingly.

$\mathbf{T}_1$ stores the IP addresses of the local controller's transceivers in each and every microgrid. Once a request is received, the SDN controller will search $\mathbf{T}_1$ to find the identification (ID) number of the microgrid. If the microgrid ID is i, the SDN

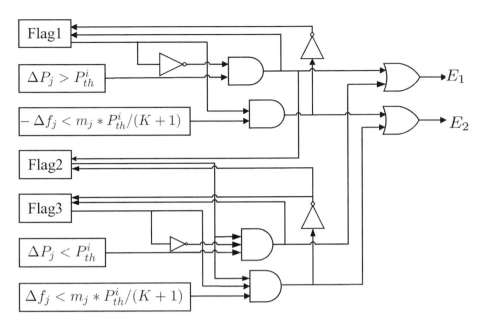

Figure 4.9 Digital circuit diagram for event detection logic (initial values of all flags are set as zero).

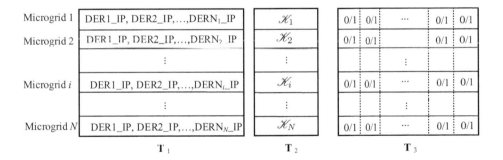

Figure 4.10 IP address table ($\mathbf{T}_1$), K-nearest neighbor table ($\mathbf{T}_2$), and communication connectivity table ($\mathbf{T}_3$) in an SDN controller.

controller will then search the ith cell in $\mathbf{T}_2$, which contains an array of the K-nearest neighbors of microgrid i ranked by electrical distances. The ith element in $\mathbf{T}_2$ is an index set $\mathscr{K}_i$, pointing to the K microgrids connected to microgrid i with shorter electrical distances. $\mathbf{T}_3$ is an $N \times N$ (0, 1)-matrix initially set as $\mathbf{0}$, meaning initially no global communication is established. As communications between microgrids are assumed to be bidirectional, the use of microgrid connectivity table $\mathbf{T}_3$ can avoid duplicated operations in the communication network. For instance, if there is an E_1 request received from microgrid i, the SDN controller would check the ith row of $\mathbf{T}_3$ and find all zero elements among $\{i,k\}, k \in \mathscr{K}_i$. If those elements on their symmetric positions are also zeros, this indicates communication link does not exist between

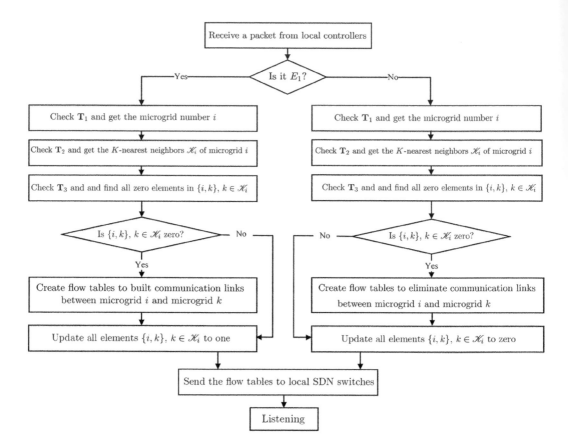

Figure 4.11 SDN-enabled event-triggered communication in networked microgrids.

microgrids i and k. The SDN controller can then generate instructions for the SDN switches to establish the links and change all zero elements among $\{i,k\}, k \in \mathcal{K}_i$ in $\mathbf{T}_3$ to ones. Similarly, if an E_2 request is received, the SDN controller finds all nonzero elements among $\{i,k\}, k \in \mathcal{K}_i$ and verifies if the elements on their symmetric positions are zeros to ensure that the links are not requested by other events. The SDN controller will then remove these links (if they exist) and update all non-zero elements among $\{i,k\}, k \in \mathcal{K}_i$ to zeros. Figure 4.11 shows this SDN-controlled process.

Event-triggered communication for microgrids is empowered by SDN and implemented directly in the cybernetwork. This differentiates it from the state of practice for most event-triggered approaches for average consensus based controllers, which are realized on the local controllers. For the SDN-enabled approach, the SDN switch captures data packets from the microgrid's leader DER under power deficiency/recovery and then forwards them to its neighboring microgrids. In contrast, the conventional event-triggered approach uses local controllers to broadcast data to their neighboring microgrids whenever events are detected. SDN for event-triggered communication is advantageous in that (i) it uses fast programmable networking instead of using controller-to-switch bandwidth, meaning it does not create communication bottlenecks on the network's edge; (ii) it enables on-the-fly generation of K-nearest neighbors for microgrids, which is self-maintained in $\mathbf{T}_2$ in the SDN controller; and (iii) it supports the implementation of NMs in a real-world complex communication network where individual microgrids use separate subnets, which makes it highly scalable and supportive of the plug-and-play of microgrids.

4.5   The Cyberphysical Networked Microgrids Testbed

A cyberphysical hardware-in-the-loop (HIL) NM testbed is designed and established to test and validate SDN-enabled distributed power sharing in NMs. This section introduces the overall architecture of the cyberphysical NM testbed, and it discusses hardware design, NM models, and the setup of the SDN cyber network.

4.5.1   Architecture of the Cyberphysical Networked Microgrids Testbed

This version of the cyberphysical NM testbed uses an OPAL-RT HIL simulator on which NMs are simulated in real time. To enable a fully software-defined, hardware-independent cybernetwork, all major cyberfunctions (e.g., event detection) and communication modules are operated on multiple virtual machines (VMs) implemented on three computing servers. A software-defined network is used to implement and manage data transmission in microgrids.

The OPAL-RT simulator hosts a library of high-fidelity microgrid DER models and inverter models. The team has also extended the simulator's functionality by incorporating cybernetworks, which has been achieved by interfacing with either communication network simulators (cosimulation) [47] or real communication hardware [25] through RT-LAB's asynchronous Ethernet blocks. This chapter adopts the latter approach for building the testbed. A real SDN backbone network is established to process data and interface with the HIL simulator, which allows us to examine the impacts of latency, failures, and anomalies in the cyber network on microgrid responses. A variety of cyberattacks on NM communication and control can also be investigated on this setting [42]. The hardware SDN is used to interconnect the real communication and computing environment with the HIL simulator to enable event-triggered data flow management, which can hardly be achieved by a traditional non-OpenFlow switch.

To achieve the high-fidelity emulation of field data flows through the SDN network, the data traffic for NM operations and power sharing should really pass through the hardware SDN switches. However, if the data are transmitted within the OPAL-RT simulator only, no data can be sent out to the SDN network. To solve this problem, we created a number of VMs, each of which has separate IP and media access control (MAC) addresses, to emulate the cybermodules in a real NM system. Those VMs are designed to be able to exchange data among each other, transmit data to and the HIL simulator. In fact, the VMs represent DERs in the NMs, which generate data to be used by the secondary control of microgrids and NMs.

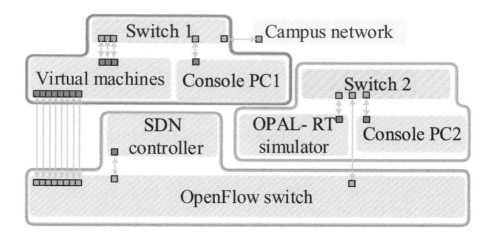

Figure 4.12 Overall structure of the cyberphysical networked microgrids testbed.

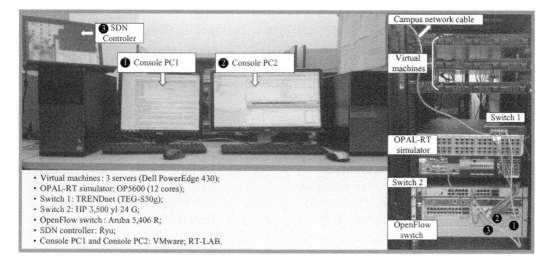

Figure 4.13 The cyberphysical HIL testbed and its components.

To summarize, this particular cyberphysical NM testbed is composed of three subsystems serving different functions:

• A real-time NM simulator and peripheral devices including a console PC and a network switch
• Three servers hosting eight VMs, which correspond to eight DERs in the test system, and auxiliary devices including a console PC and a network switch
• An SDN network including an SDN controller and OpenFlow switches

The software-defined network interconnects the three subsystems, which share a private Ethernet subnet (see Figure 4.12). Each server provides a management port that links to the campus network. This will allow the console PCs to remotely operate all VMs. Figure 4.13 shows the cyberphysical testbed for NM research.

4.5.2 The Cyberphysical Simulator and Networked Microgrids Model

An Opal-RT OP5600 simulator with 12 enabled cores is the major component of this testbed, which can perform electromagnetic transients level simulations in real time. As it is equipped with a gigabit Ethernet port, this simulator is able to communicate with the console PC2 through TCP/IP while it talks with the VMs through UDP/IP. Console PC2 serves as a workstation where the NM models can be built, edited, and compiled through RT-LAB and sometimes directly with MATLAB toolboxes. To avoid simulation overrun, the NM is partitioned into eight subsystems simulated on eight cores in parallel; moreover, the ARTEMiS-SSN technique is activated to enforce a fixed time-step nodal admittance solver to further speed up the simulation [48].

Data exchanges in power-sharing processes are enabled by multiple UDP sockets allocated to two Opal-RT cores, one for sending and another for receiving from the VMs. Inside the simulator, the shared memory allows the synchronization of data exchange with the NM simulation [25]. The dedicated simulator cores perform data fusion operations that assemble DER measurements and control signals (e.g., frequency f_j, active power P_j, and secondary control signal f'_j) into one packet. The fused data are then transmitted to corresponding VMs, where event detection or data exchange will take place.

Three Dell PowerEdge R430 servers are used to run the VMs. There are a total of 12 Ethernet ports on the servers, one of which is dedicated to connecting with console PC1. Eight of the remaining ports are interfaced with eight VMs one by one. On each VM runs four instances of the data transceiver program that establish four UDP sockets to enable communications with the simulator and any other VMs. The programs that realize the event detection logic (see Figure 4.9) also run on the VMs, which generates E_1 or E_2 signals and then transmits them to the SDN controller. Based on the nature of events, the SDN controller then establishes or eliminates communication links as required.

Based on this testbed, a four-microgrid NM test system has been built. Each microgrid consists of two DERs, which are regulated by VSIs and supply two local customer loads. Those VSIs are controlled via pulse-width modulation (PWM) as seen in Figure 4.8. In each microgrid, the power generated from DERs flows into the NM backbone network through a local PCC bus (see Figure 4.14). Section 4.6 provides the cyberphysical parameters of this test system.

4.5.3 Inside the Networked Microgrid Model

This subsection provides an overview for the RT-LAB model of the NM system to offer an inside look at the Simulink-like model structures. A step-by-step tutorial for setting up the simulation environment is also introduced.

First, a console PC with MATLAB and RT-LAB installed should be connected with the simulator. In this NM model, for instance, the IP address of the target simulator is set as 192.168.10.101. Meanwhile, the IP address of the console PC should be in the same subnet with the target simulator; for instance, it can be set as 192.168.10.102.

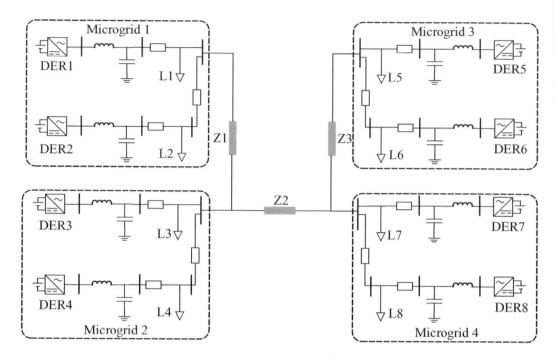

Figure 4.14 The structure of the networked microgrids.

To execute the simulation, RT-LAB is started, and the graphical user interface (GUI) will pop up as shown in Figure 4.15. To load and run the NM model, readers can follow the following steps:

1. Create a new project or import existing projects into the workspace as shown in Figures 4.16 and 4.17.
2. Click on Open Model and then navigate through local files to select the system model (.mdl format).
3. Choose a target simulator and compile the model by clicking on Compile in the Main Control Panel. The following steps will be executed:
   (i) Partition the test system into subsystems.
   (ii) Generate C codes for simulating subsystems.
   (iii) Compile the C codes for subsystems and build executable files that can run on Opal-RT.
4. Allocate executable files to simulator cores by clicking on the Assign button. All executable files for subsystems and available computing cores show up; then one can assign each and every executable file to a simulator core
5. Click Load to upload the executable files: If Load is checked, it means that the RT-LAB program can start successfully; otherwise, RT-LAB will display diagnostic information (one can click Reset to retry the model)
6. Click on Execute to start the real-time simulation.

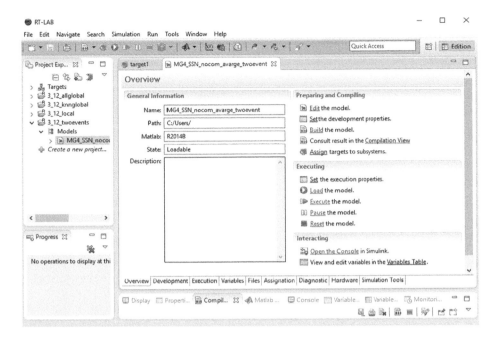

Figure 4.15 User interface (UI) of simulation software.

Figure 4.16 Import an existing project.

Model of Distributed Power Sharing in Networked Microgrids

First, the overall structure for distributed power sharing in NMs is summarized in Figure 4.18, where the NM system consists of microgrids along with a control model, console module, and real-time simulation module.

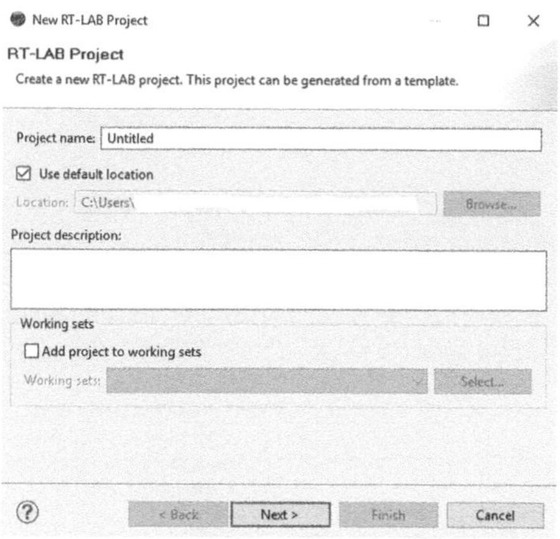

Figure 4.17 Create a new project.

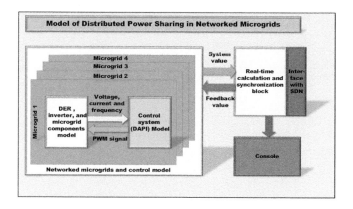

Figure 4.18 Structure of distributed power sharing in NMs.

An overview of the NM model built in the Opal-RT simulator is shown in Figure 4.19. The NM system consists of three components, that is, $SM\_Control$, $SS\_Network$, and $SC\_Console$, as shown in Figure 4.19.

The $SM\_Control$ block coordinates the real-time calculation and synchronization of the NM system. As shown in Figure 4.20, this system enables I/O communications, providing interfaces with SDN and microgrid controllers VMs. For $Message1$ through 8, each message contains the frequency, voltage, active power, and reactive power outputs for one DER. $Con1$ through 8 transfer the frequency of the DERs and the control signals used for droop and DAPI control of the NM system. The subsystem in $SM\_Control$ is shown in Figure 4.21, where $OpWriteFile$ is a module that saves the operating data into files.

The SS_Com module is shown in Figure 4.22. Take DER **1** as an example: there is a pair of $AsynchronousRec$ and $AsynchronousSend$ modules for DER **1**. The $AsynchronousSend$ module sends out the variables (e.g., voltage, current, and

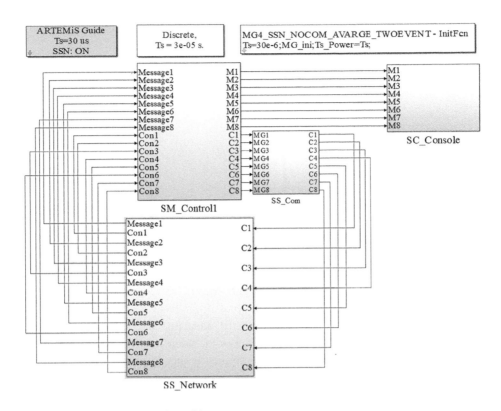

Figure 4.19 The networked microgrids system.

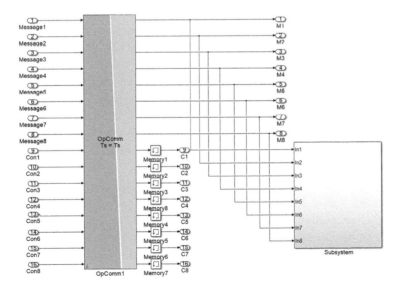

Figure 4.20 SM_Control.

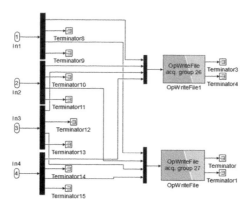

Figure 4.21 The subsystem in SM_Control.

frequency) of DER **1** to the neighboring DERs and microgrids. The *Asynchronous Rec* module receives the variables from local neighboring DERs. The module *OpIPSocketCtrl1* configures the IP addresses of this pair of modules. Another *AsynchronousRec* module receives the variables from the neighboring microgrids. Module *OpIpSocketCtrl9* configures the listening port for the data packets from neighboring microgrids.

Messages 1 through 8 can be exported through link $M1$ through $M8$ and observed in *SC_Console* (see Figure 4.23), which allows for run-time interactions between the user and the NM system.

SS_Network includes the NM model and control algorithms. The networked microgrid test case consists of four microgrids, as shown in Figures 4.24 and 4.25.

In Figures 4.26 and 4.27, $f1$ is the frequency of DER **1** and $dF1$ is the secondary frequency control variable f_1'; $V01$ and $I01$ are the voltage and current of $VI2$ separately. $IL1$ is the local current of $VI1$. $C1$ includes the frequency and secondary frequency control variables for the neighboring microgrids.

As discussed in Subsection 4.3.1, microgrids share information through their leader DERs' transceivers. For example, DER **1** is selected as the leader in Microgrid 1, while the DER 2 is the follower. Unlike the follower DER, the leader is equipped with global power-sharing capability.

Figure 4.28 shows the control system of DER **1**, which consists of four modules. The *Primary and Secondary Control* module realizes the droop and distributed-averaging proportional-integral control algorithms (see Figures 4.29 and 4.30). It should be noted that the voltage control loop is embedded in this module. *abc_dq* conducts Park's transformation from *abc* to *dq* coordinates. *Current Control* refers to the current control loop that generates the PWM reference signal for inverter control.

For a follower DER, such as DER 2 in Microgrid 1, the power-sharing control can be modeled as shown in Figure 4.31. It should be noted that a follower DER has no global power-sharing control in this model. Figures 4.32 and 4.33 show the Park Transformation block and PWM Generator block, respectively.

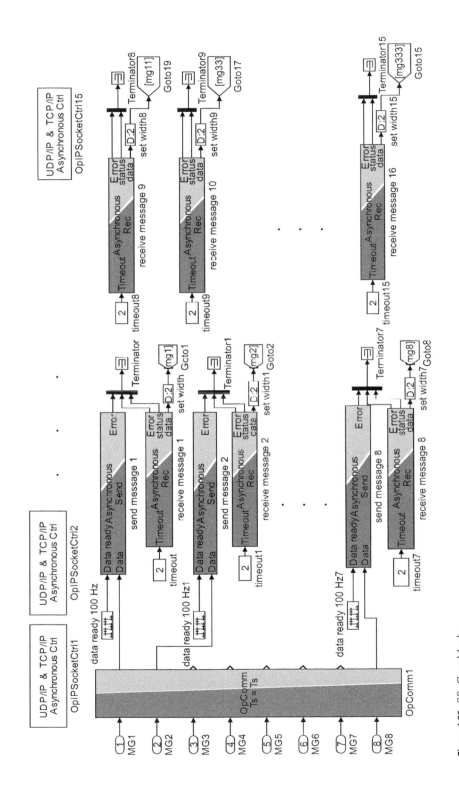

Figure 4.22 SS_Com block.

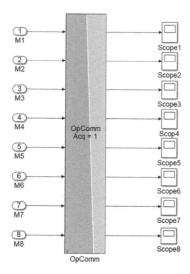

Figure 4.23 SC_Console.

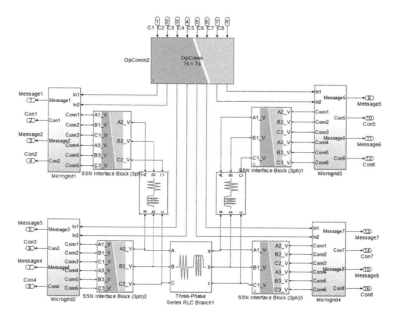

Figure 4.24 SS_Network.

4.5.4   Event-Triggered Communication through SDN

The SDN network of the cyberphysical NM testbed consists of an Aruba 5406R SDN switch supporting the OpenFlow protocol and an SDN controller. The Aruba 5406R switch features ultralow latency (1,000 Mb latency <2.8 μs) and a high throughput (up to 571.4 Mpps) [49]. For the test NM system, eight VMs are created for the eight DERs. Therefore, eight ports on the OpenFlow switch are used to connect with those

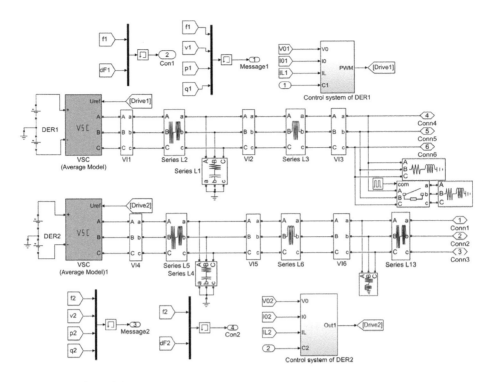

Figure 4.25 Microgrid 1.

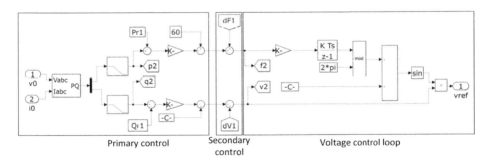

Primary control   Secondary control   Voltage control loop

Figure 4.26 Primary and secondary control.

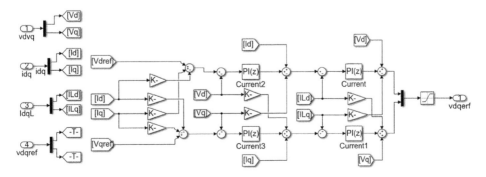

Figure 4.27 Current control loop.

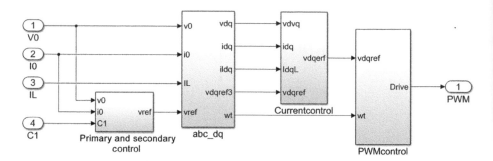

Figure 4.28 Control system of DER **1**.

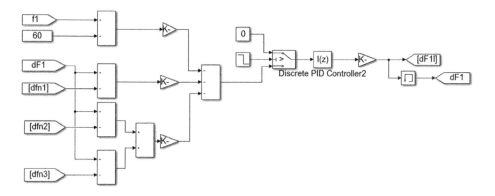

Figure 4.29 Secondary frequency control block.

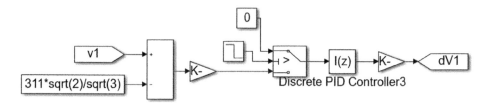

Figure 4.30 Secondary voltage control block.

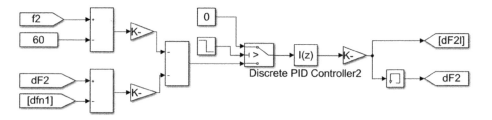

Figure 4.31 Local power-sharing control of DER 2.

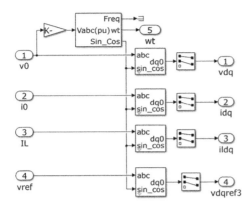

Figure 4.32 *abc* to *dq* transformation.

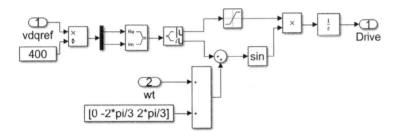

Figure 4.33 PWM control signal block.

VMs. The ninth port is used to link to the HIL simulator while another port is assigned for the SDN controller. Microgrid and NM power sharing as well as other advanced functions are based on a Ryu SDN controller, which allows fast development of those SDN applications.

Our NM power-sharing application will create an IP address table $\mathbf{T}_1$ in the SDN controller. For our four-microgrid NM system, a two-nearest neighbors table $\mathbf{T}_2$ is generated by examining the AC backbone of the NMs. By searching the communication connectivity table ($\mathbf{T}_3$), new flow rules are automatically generated by the SDN controller and thus data paths are created or canceled out through a pipeline supported by OpenFlow.

The flow entries in flow tables define the operation rules of the OpenFlow switch. When a packet arrives, the matching fields in it, such as the source IP address, will be used to find matched entries in the flow table. Once it finds a matched entry, the SDN switch will execute those instructions written in that entry. In an OpenFlow switch, a pipeline of flow tables (see Figure 4.34) is the container that predefines the instruction sets for the packets. Those flow tables in a pipeline operate in a forward-only order. For the event-triggered microgrid communication, the SDN controller will generate instructions written in the flow tables. In particular, the SDN controller performs the control functions illustrated in Figure 4.10, updating the switch automatically according to changes in the NM system.

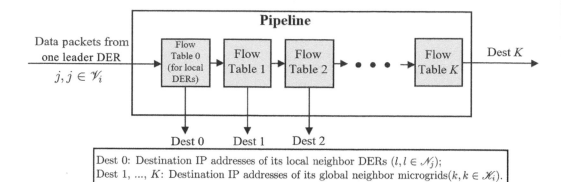

Dest 0: Destination IP addresses of its local neighbor DERs $(l, l \in \mathcal{N}_j)$;
Dest 1, ..., K: Destination IP addresses of its global neighbor microgrids$(k, k \in \mathcal{K}_i)$.

Figure 4.34 Pipeline processing of NM data on the SDN switch.

To give an example, let us assume an E_1 request is received by the SDN controller. Immediately, new flow rules will be established by the SDN controller for flow tables 1 to K and then the updated rules will be sent out to and activated in the OpenFlow switch. As a result, data packets for power sharing are able to be forwarded to the K neighboring microgrids that can be reached by their IP addresses. When a packet arrives at the switch, it will match the rules in flow table 0, and the OpenFlow switch will forward it to its local DER (Dest 0) based on the header field of the matched packet. Meanwhile, the packet will also be forwarded to the next flow table where the destination IP in the header field will be changed to the IP address of a neighboring microgrid chosen for fulfilling the global power sharing. Since we can choose 0 to K microgrids for power sharing depending on customers' requirements, the packet forwarding process will eventually lead to the establishment of a maximum of $2 \times K$ new links for bidirectional communications needed for the NM power sharing. In our particular test case, if DER1 (Microgrid 1) initiates the E_1 signal, the packets being sent from DER1 to DER2 will be broadcast to two neighboring microgrids as defined in $\mathbf{T}_2$, that is, DER3 (Microgrid 2) and DER7 (Microgrid 4).

Even though the event-triggered communication operations can be realized by either a hardware SDN switch or a software one, we choose to use a hardware OpenFlow switch to implement the pipeline model and packet header modification. The benefit is that very little overhead (on the microsecond level) will be introduced at the hardware level of the OpenFlow switch, which is highly desirable for resilient microgrid and NM operations.

4.6   Testing and Validation

The effectiveness of the SDN-based NMs' communication and distributed control strategy is validated through a few comprehensive tests performed on our cyberphysical HIL testbed. The overarching goal of those tests is to examine whether the

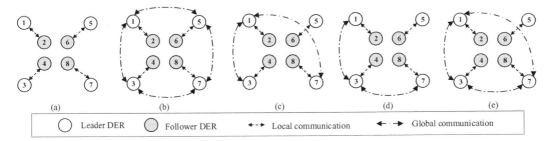

(a)　　　　　(b)　　　　　(c)　　　　　(d)　　　　　(e)

○ Leader DER　　◉ Follower DER　　◄- -► Local communication　　◄- ·-► Global communication

Figure 4.35 SDN-enabled network topologies for the NM system: (a) local communication for each microgrid (baseline communication for all cases); (b) looped global communication involving all leader DERs; (c) star global communication centered at DER1; (d) star global communication centered around DER3; and (e) global communication requested concurrently by DER1 and DER3.

integration of global power sharing, K-nearest neighbor microgrid clustering, and SDN-empowered event-triggered communication could lead to superior performance in terms of communication cost and system response. First, several single-event cases are designed to verify the NM control performances. Further, a few multiple-event cases are studied to test the resilience of the SDN controller impacted by a series of events, which represents those rare but high-impact risks in real NMs.

Two types of studies are designed to test the NM control and communication performance based on different scenarios of microgrid power deficiency. *Study I*, which is elaborated in Section 4.6.1, focuses on a single-event case: a step change of load L_1 (see Figure 4.14) in Microgrid 1 from 10 kW to 15 kW, which triggers an E_1 request from DER1. *Study II*, detailed in Section 4.6.2, includes two double-contingency cases: the first case includes two independent contingencies with L_1 and L_3 increasing by 5 kW at two different time instances without causing communication overlap, and another one with the same magnitudes of L_1 and L_3 changes occurring at very close time instances, which tests the smartness of the SDN controller in dealing with such complex issues.

The following simulation setting applies for all test cases: a simulation time step of $30\mu s$, a sampling rate of 33.333 kHz for data communication and visualization, and a simulation time of 60 s. Table 4.1 gives the electrical and control parameters for the test NMs. The droop control and the DAPI secondary control are coordinated in such a way that, after the droop control reaches steady state for one second, the secondary control will kick in. In the first double-contingency case, L_1 increases at 6 s and recovers at 36 s while L_3 increases at 18 s and is recovered at 48 s. In another double-contingency case, immediately after L_1 increases at 6 s, L_3 ramps up at 6.6 s and reduced back to the original level at 36.6 s. The detection threshold P_{th} is chosen to be 20% of the power rating of each microgrid, which is 3 kW. The number of neighboring microgrids K is set as 2 for the test NM system. Figure 4.35 shows the communication network topologies adopted in the case studies.

Table 4.1 Four-microgrid NM system parameters.

Parameter	Value
Nominal frequency	60 Hz
DC voltage	800 V
Nominal voltage (line–line RMS)	311 V
Filter inductance	1.35 mH
Filter capacitance	50 μF
Line impedance Z_1	R = 1 ohm, L = 10mH
Line impedance Z_2	R = 5 ohm, L = 20mH
Line impedance Z_3	R = 1 ohm, L = 10mH
–	Leaders DERs – Follower DERs (1,3,5,7) – (2,4,6,8)
Rated active power	10 kW – 5 kW
Rated reactive power	5 kVar – 2 kVar
f-P droop coefficients	0.6e-5 Hz/W – 1.2e-5 Hz/W
V-Q droop coefficients	1.2e-3 V/Var – 2.4e-3 V/Var
Frequency restoration coefficients	10 – 10
Local power-sharing coefficients	10 – 10
Global power-sharing coefficients	100 – 100

4.6.1   Study I: The Single-Event Scenario

A. Effectiveness of global power sharing

To verify the need for global power sharing, we compare the NM responses where there exists only the local power sharing (see Figure 4.36a) versus the NM responses where there exists global power-sharing control in the NMs (see Figure 4.36b). To further illustrate the performance of the hierarchical power-sharing scheme, Figure 4.37 shows the time-domain curves of those secondary frequency control signals for the two cases.

The following observations can be made:

- Figure 4.36a shows that DERs within any one of the microgrids are able to communicate through SDN to trigger the local distributed power sharing based on DAPI. It can be seen that DER1 has a power spike around 13.5 kW initially, which is reduced to approximately 12.5 kW due to the local power sharing within Microgrid 1. The total extra power generation from Microgrid 1 (including DER1 and DER2) is 3.7 kW, meaning only a small portion of the load increment (1.3 kW) is supplied by the three other microgrids. As a result, the total power output of Microgrid 1 exceeds 120% of its nominal capacity. This indicates an emergency state of Microgrid 1 in that it has likely run out of its reserve and cannot support any further load increase. In practice, the load shedding protection may be activated to bring Microgrid 1 back to a predefined secure operation region with an adequate reserve.
- In the case illustrated by Figure 4.36b, bidirectional looped communication is established for the four leader DERs in those microgrids. This allows the global

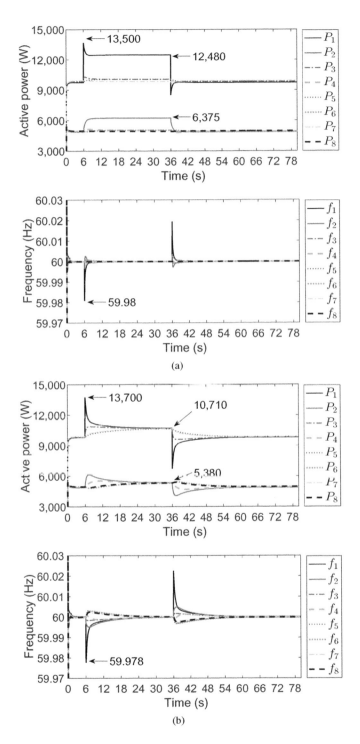

Figure 4.36 NM performances. (a) Local power sharing only corresponding to Figure 4.35a. (b) Global power sharing empowered with looped communication corresponding to Figure 4.35b. The steady-state power-sharing results (DER1 and DER2 power outputs and system frequency) before load recovery are labeled on both Figures 4.36 and 4.38.

power sharing among the four microgrids. Subsequently, all DERs in each and every microgrid will achieve proportional power sharing at the local microgrid level. Figure 4.36b shows that, before the load recovery occurs at 36 s, Microgrid 1 only needs to contribute less than one fourth of the total demand surge, that is, the two DERs in Microgrid 1 just need to increase their outputs by 1.1 kW in total. In general, the global power-sharing scheme would help distribute the extra load evenly among those networked microgrids, meaning the burden on each microgrid would less likely become an emergency. Even if any microgrid may run into an emergency state, it would be less severe and more correctable.

- Figure 4.37a shows that only f_1' of DER1 and f_2' of DER2 can respond to the demand change relatively quickly when only local power sharing is equipped in microgrids. The load increase in L1 is shared locally between DER1 and DER2, where a consensus can be reached efficiently. In contrast, Figure 4.37b shows a slower consensus process when the global power sharing is introduced: it takes the leader DERs about 1 s to ramp up their power outputs whereas it takes the follower DERs up to nearly 20 s to catch up with the leaders. By examining the two-layer scheme in Figure 4.8, it can be seen that first a global consensus algorithm transmits the control signals to the leader DERs, and then the follower DERs start a local consensus process. Eventually, the two-layered power sharing can eventually allocate the increased loads proportionally to all the DERs.

- An interesting phenomenon revealed in Figure 4.37b is that Microgrid 2 (f_3', f_4') responds more rapidly to the load increase in Microgrid 1 than Microgrids 3 and 4 do. This verifies the statement from Subsection 4.3.2 that those microgrids strongly connected would reach a consensus faster than those loosely connected. This means that we could further improve the performance of two-layered NM power sharing by exploiting microgrid neighborhood sets, which will be demonstrated next.

B. Efficacy of K-nearest-neighbor-based microgrid clustering

The test results have also verified the efficacy of the K-nearest neighbor approach to microgrid clustering. Once Microgrid 1's two-nearest neighbors are obtained by the electrical distance matrix, only two microgrids would participate in the global power sharing process. Accordingly, the looped communication network involving all microgrids is reduced to a star network centered around Microgrid 1. The following can be observed:

- A comparison of Figure 4.38a with Figure 4.36b indicates that the two-nearest neighbor clustering (Microgrids 1, 2, and 4) performs equally well as the all-neighbor clustering with a reduced communication complexity. Please note that those microgrids excluded from communications are not fully stopped from contributing power to the microgrid(s) in need. Kirchhoff's law always works within electrically interconnected microgrids. However, the amount of power they could contribute would be very small or negligible due to the large electrical distances.

- The continuous communications among the K-nearest neighbors can be upgraded into a discrete-event-triggered communication, which will further improve the

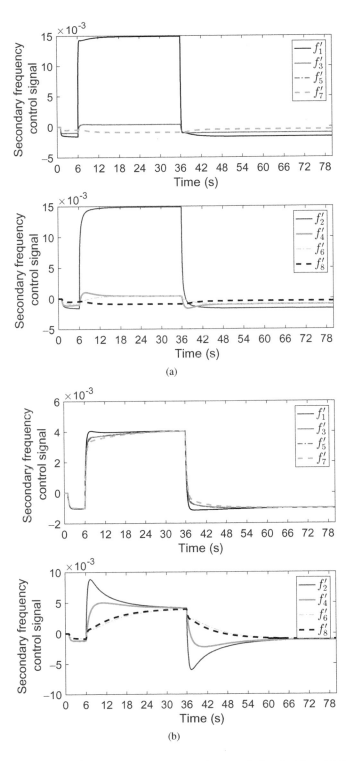

Figure 4.37 f_j' signals for leader DERs (upper subplots in (a) and (b)) and follower DERs (lower subplots in (a) and (b)): (a) Local power sharing corresponding to Figure 4.36a; (b) Global power-sharing corresponding to Figure 4.36b.

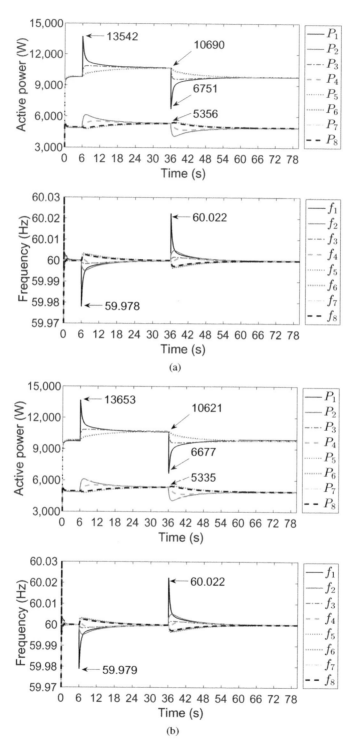

Figure 4.38 DER power and system frequency with a two-nearest-neighbors-based global power sharing (Figure 4.35c) in response to MG1 power shortage. (a) Continuous communication; (b) event-triggered communication.

system efficiency. The advantage of the event-triggered scheme can be seen in Figure 4.38b, where the event-triggered power sharing for Microgrid 1 is the same effective as the continuous one. There are slight differences between the results of the two schemes, as shown in Figure 4.38b. In the event-triggered solution, DER1 stops changing its output once the tolerance Δf_j^{th} is satisfied (the secondary frequency control signal no longer updates). With the continuous communication scheme, DER1 continues updating output till when Δf_j approaches zero. If the NM customer would like to eliminate the differences, one can simply modify the event-triggered controller code so that when an E_2 request is received, rather than immediately closing out the global communication, sparse data packets can still be transmitted to allow further reducing the frequency error.

- The control signals recorded during the event-triggered power sharing and the power-sharing process under continuous communication are compared in Figure 4.39. The contributions of different control signals to (f_1') (secondary frequency control signal) are illustrated by the shaded areas, where the positive area means the weakening effect. During the process, it can be seen that the frequency droop and local DAPI signals try to stimulate f_1' after the load increase occurs; meanwhile, the global consensus algorithm averages f_1' with $f_i', i \in \mathscr{K}_i$ (the leader DERs in the $\mathscr{K}$-neighbor microgrids), which counters the change.

- Figure 4.39a,b shows that the event-triggered communication causes an instantaneous rise when a discrete event occurs. Nonetheless, the secondary frequency control variable in steady state was not compromised by the initial transients. A desirable feature of the event-triggered communication is that, once the global communication is ended, no new f_i' signal will be fed to the local controllers, which effectively stops those microgrids neighbors from continuing to adjust power exchanges once an NM emergency is dismissed.

In summary, K-nearest-neighbor microgrid clusters are suitable for event-triggered communication as they do not compromise the power-sharing performance. Rather, they enable fast power support among NMs and reduce communication costs.

C. Advantages of SDN-empowered event-triggered communication

A major innovation in this chapter is that the event-triggered NM communication is enabled by SDN. This case study demonstrates particularly the performance of SDN. As introduced before, the control and communication of each DER are modeled in a VM. Thus a VM is like a "digital twin" of a DER, and the two terms are used interchangeably in this chapter. Similarly, the HIL NM simulator is a digital twin of the physical infrastructure of an NM system, and we may use the HIL simulator and the NM system interchangeably. Each VM has four UDP sockets set up for communication purposes: Socket1 receives signals from the NM system and broadcast them to its locally connected VMs/DERs; Dual to Socket1, Socket2 receives signals from locally connected VMs and transmit them back to the NMs; and Socket3 and Socket4 are designed to receive data from the OpenFlow pipeline where the the SDN controller provides on-the-fly configuration and update for the flow tables. Figure 4.40a exemplifies the data flows of VM1, where the potential data flow required

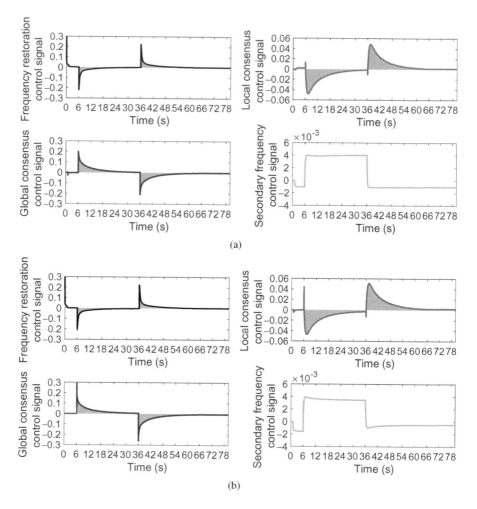

Figure 4.39 DER1 control signals. (a) Continuous communication corresponding to Figure 4.38a; (b) Event-triggered communication corresponding to Figure 4.38b. Upper-left subplot: frequency restoration control signal. Upper-right subplot: local consensus control signal. Lower-left subplot: global consensus control signal. Lower-right subplot: secondary frequency control variable.

by local power sharing (see Figure 4.35a) is illustrated by the solid lines and those for global power sharing (see Figure 4.35b–e) by the dashed lines.

To help understand what happens in the local data traffic when an event-triggered power sharing in Figure 4.38b is performed, we use Wireshark, a popular network monitoring tool [50], to collect all packets in VM1 every 100 ms and the recorded data throughput of VM1 can be seen in Figure 4.40b. At 6 s, a power shortage emergency in Microgrid 1 is captured by the detection function on Socket1 of VM1, and subsequently an E_1 request is initiated. Upon receiving the request, the SDN controller searches T_1 and T_2 and pinpoints the two neighbors of Microgrid 1, that is, Microgrids 2 and 4. Next, two flow rules that add two data flows DER1 $\longleftrightarrow$ DER3 and DER1 $\longleftrightarrow$ DER7 (Figure 4.35c) are created in the pipeline flow table. Right

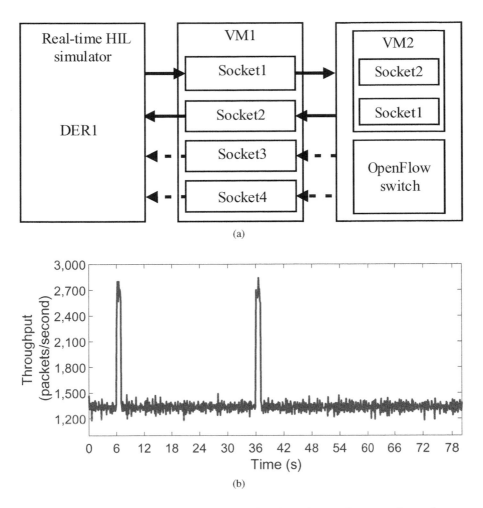

Figure 4.40 (a) VM1/DER1 data flows; (b) VM1 data throughput under event-triggered power sharing.

after this, new traffic for information sharing among Microgrid 1 and its two nearest neighbors are enabled by the OpenFlow switch. Due to the utilization of the new links during the global power-sharing process, it can be seen in Figure 4.38b that the VM1 throughput is doubled during $6\,s \sim 7\,s$ and $36\,s \sim 37\,s$. Later, an E_2 signal is received, suggesting that the required power sharing is successfully accomplished, the SDN controller will cancel out those newly established communication links. Figure 4.38b also clearly shows that only a very short duration ($1\,s$ for each contingency) of global traffic is needed for the event-triggered scheme, which has minimized the bandwidth usage compared to the traditional continuous communication scheme.

4.6.2 Study II: Multiple-Contingency Cases

Two double-contingency cases, with and without communication overlap, are used to test the performance of the SDN controller. In Figure 4.41a, $6\,s \sim 7\,s$, $18\,s \sim 20\,s$,

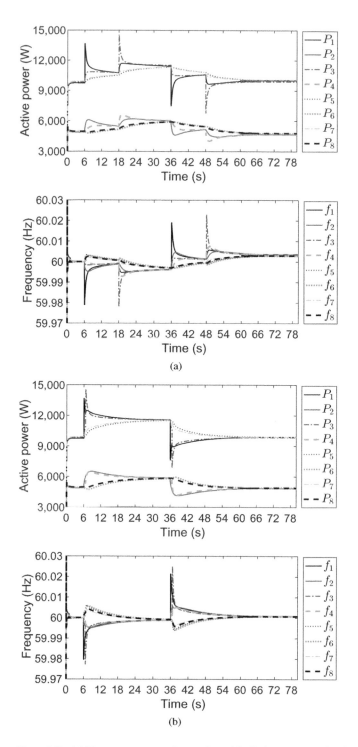

Figure 4.41 (a) Two separate contingencies with distinct communication schemes corresponding to Figure 4.35c, d; two overlapped contingencies with shared communication links corresponding to Figure 4.35e.

36 s $\sim$ 37 s, and 48 s $\sim$ 50 s are the time intervals with global communication. DER1 and DER3 send E_1 requests at 6 s and 18 s and the global power sharing for the two microgrids finishes in 1 s and 2 s, respectively.

Even though the chance to have double power deficiency contingencies within one second is low, the impact of such double- or higher-order contingencies is much higher, resulting in considerable risks. Therefore, it is necessary to examine the capability of the SDN controller in handling rare events. As shown in Figure 4.41b, right after L_1 increases at 6 s, L_3 increases at 6.6 s, initiating global communication in Microgrid 2, and expands the global power sharing of Microgrid 1–7.7 s. The contribution of each microgrid is about 2.5 kW to make up an increased demand of 10 kW during the steady state of the power-sharing process.

The microgrid data flows during the event-triggered power-sharing process are examined to understand the SDN performance in response to the double-contingency cases. The new communication links established in Microgrid 1 are DER1$\longleftrightarrow$DER3 and DER1$\longleftrightarrow$DER7 (see Figure 4.35c) while those requested by Microgrid 2 are DER1$\longleftrightarrow$DER3 and DER3$\longleftrightarrow$DER7 (see Figure 4.35d). Thus, two global links will be established for DER1 to handle the first contingency while only one link is needed for DER1 in the second event. It can be seen in Figure 4.42a that, for the two separate contingencies, the VM1 (DER1) throughput is doubled when Microgrid 1 sends requests to the SDN controller during 6 s $\sim$ 7 s and 36 s $\sim$ 37 s, whereas there is only 50% increase in the throughput when Microgrid 2 sends request during 18 s $\sim$ 20 s and 48 s $\sim$ 50 s.

For the double-contingency case with overlapped communications (see Figure 4.42b), the requests sent for the discrete events include (1) $E1$ by VM1/DER1, (2) $E1$ by VM3/DER3, (3) $E2$ by VM1/DER1, (4) $E2$ by VM3/DER3, (5) $E1$ by VM1/DER1, (6) $E1$ by VM3//DER3, (7) $E2$ by VM3/DER3, and (8) $E2$ by VM1/DER1. As Figure 4.42b shows, as the two contingencies occur nearly concurrently within 0.6 s, the doubled throughput remains unchanged during the overlapped periods 6 s (1) $\sim$ 7.7s (4), and 36 s (5) $\sim$ 37.7 s (8).

In the aforementioned experiments, the SDN controller is found capable of avoiding issues such as deadlocks, collisions due to multiple requests, and redundant operations in the SDN network. The experiments also verify the correctness of the event detection and processing design discussed in Subsection 4.4.2. This means that the SDN-based event-triggered approach is promising to cope with rare events such as double-, triple-, or higher-order contingencies in future NM operations.

4.7   Conclusion and Guide for Future Applications

The focus of this chapter is the introduction of SDN to the networking and operations of microgrids. It gives a tutorial about SDN to help readers understand SDN's programmability and how this may empower resilient networked microgrids. A first usage of SDN is to architect a layered power-sharing method for NMs, where SDN is used to architect and realize flexible and lightweight event-triggered communication and

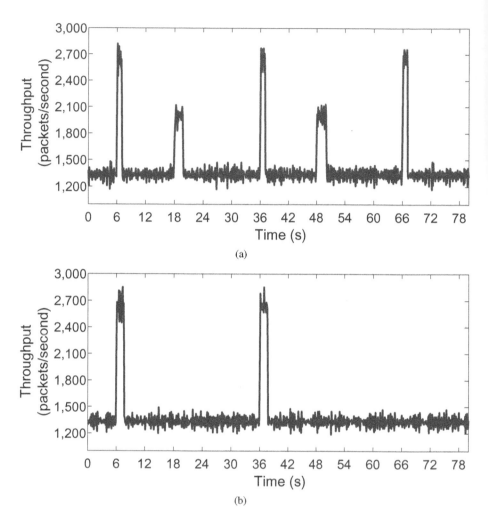

Figure 4.42 (a) VM1 data throughput during two separate contingencies; (b) VM1 data throughput during two overlapped contingencies.

control. This distributed power-sharing scheme boils down to a simple global power-sharing module added on each of the local inverter controllers as part of the leader DERs in the NM system. When the number of microgrids in an NM system is large, the concept of K-nearest-neighboring (K-NN) microgrids can be adopted to reduce the complexity of communications. The SDN controller can retrieve K-NN online by looking up the electrical distance information in the NM network. The SDN controller is particularly useful for setting up an event-triggered communication scheme such that only when major structural or operational condition changes (e.g., power outage, shortage, or load recovery) occur, the global communication mechanism is enabled. The SDN functionalities are able to achieve a highly resilient NM communication network by avoiding congestion in both the backbone communication networks and controller-to-switch data paths. It also has the potential to improve the cyberresilience of NMs by implementing cyberattack detection, localization, and mitigation techniques on the SDN controllers.

It is relatively amendable to realize the SDN-based approach to networking and operating microgrids. As a guide, three modifications may need to be made on the existing microgrid infrastructures to achieve the global power sharing among networked microgrids:

- For specific leader DERs, a global power-sharing and communication interface are to be implemented (see Figure 4.8).
- For DERs in microgrids, the event detection function is to be implemented as part of the inverter control (see Figure 4.9).
- Use a number of SDN switches to elevate the existing communication infrastructure to a scalable SDN network.

Normally the inverter controllers for microgrid DERs have the ability to support extensibility so that the modules for the first two modifications can be added. The transformation to an SDN network is also manageable through the use of either virtual or hardware SDN switches. Given the fast development of SDN solutions, either way is now affordable and efficient.

References

[1] R. H. Lasseter, "Smart Distribution: Coupled Microgrids," *Proceedings of the IEEE*, vol. 99, no. 6, pp. 1074–1082, 2011.

[2] F. Shahnia, R. P. Chandrasena, S. Rajakaruna, and A. Ghosh, "Autonomous Operation of Multiple Interconnected Microgrids with Self-Healing Capability," in *2013 IEEE Power and Energy Society General Meeting (PES)*, 2013, pp. 1–5.

[3] M. J. Hossain, M. A. Mahmud, F. Milano, S. Bacha, and A. Hably, "Design of Robust Distributed Control for Interconnected Microgrids," *IEEE Transactions on Smart Grid*, vol. 7, no. 6, pp. 2724–2735, 2016.

[4] X. Fang, Q. Yang, J. Wang, and W. Yan, "Coordinated Dispatch in Multiple Cooperative Autonomous Islanded Microgrids," *Applied Energy*, vol. 162, pp. 40–48, 2016.

[5] Z. Wang, B. Chen, J. Wang, and C. Chen, "Networked Microgrids for Self-Healing Power Systems," *IEEE Transactions on Smart Grid*, vol. 7, no. 1, pp. 310–319, 2016.

[6] J. Li, X.-Y. Ma, C.-C. Liu, and K. P. Schneider, "Distribution System Restoration with Microgrids Using Spanning Tree Search," *IEEE Transactions on Power Systems*, vol. 29, no. 6, pp. 3021–3029, 2014.

[7] C. Yuen, A. Oudalov, and A. Timbus, "The Provision of Frequency Control Reserves from Multiple Microgrids," *IEEE Transactions on Industrial Electronics*, vol. 58, no. 1, pp. 173–183, 2011.

[8] T. Lv and Q. Ai, "Interactive Energy Management of Networked Microgrids-Based Active Distribution System Considering Large-Scale Integration of Renewable Energy Resources," *Applied Energy*, vol. 163, pp. 408–422, 2016.

[9] P. Kou, D. Liang, and L. Gao, "Distributed EMPC of Multiple Microgrids for Coordinated Stochastic Energy Management," *Applied Energy*, vol. 185, pp. 939–952, 2017.

[10] J. He, Y. Li, B. Liang, and C. Wang, "Inverse Power Factor Droop Control for Decentralized Power Sharing in Series-Connected Micro-Converters Based Islanding Microgrids," *IEEE Transactions on Industrial Electronics*, 2017.

[11] C. Wang, Y. Li, K. Peng, B. Hong, Z. Wu, and C. Sun, "Coordinated Optimal Design of Inverter Controllers in a Micro-Grid with Multiple Distributed Generation Units," *IEEE Transactions on Power Systems*, vol. 28, no. 3, pp. 2679–2687, 2013.

[12] V. N. Coelho, M. W. Cohen, I. M. Coelho, N. Liu, and F. G. Guimarães, "Multi-Agent Systems Applied for Energy Systems Integration: State-of-the-Art Applications and Trends in Microgrids," *Applied Energy*, vol. 187, pp. 820–832, 2017.

[13] F. Dörfler, J. W. Simpson-Porco, and F. Bullo, "Breaking the Hierarchy: Distributed Control and Economic Optimality in Microgrids," *IEEE Transactions on Control of Network Systems*, vol. 3, no. 3, pp. 241–253, 2016.

[14] J. W. Simpson-Porco, Q. Shafiee, F. Dörfler, J. C. Vasquez, J. M. Guerrero, and F. Bullo, "Secondary Frequency and Voltage Control of Islanded Microgrids via Distributed Averaging," *IEEE Transactions on Industrial Electronics*, vol. 62, no. 11, pp. 7025–7038, 2015.

[15] F. Guo, C. Wen, J. Mao, and Y.-D. Song, "Distributed Secondary Voltage and Frequency Restoration Control of Droop-Controlled Inverter-Based Microgrids," *IEEE Transactions on Industrial Electronics*, vol. 62, no. 7, pp. 4355–4364, 2015.

[16] B. Koldehofe, F. Dürr, and M. A. Tariq, "Tutorial: Event-Based Systems Meet Software-Defined Networking," in *Proceedings of the 7th ACM International Conference on Distributed Event-Based Systems*. ACM, pp. 271–280, 2013.

[17] Y. Yuan, D. Lin, R. Alur, and B. T. Loo, "Scenario-Based Programming for SDN Policies," in *Proceedings of the 11th ACM Conference on Emerging Networking Experiments and Technologies*. ACM, p. 34, 2015.

[18] J. McClurg, H. Hojjat, N. Foster, and P. Černý, "Event-Driven Network Programming," in *ACM SIGPLAN Notices*, vol. 51, no. 6. ACM, pp. 369–385, 2016.

[19] X. Wang and M. D. Lemmon, "Event-Triggering in Distributed Networked Control Systems," *IEEE Transactions on Automatic Control*, vol. 56, no. 3, pp. 586–601, 2011.

[20] G. S. Seyboth, D. V. Dimarogonas, and K. H. Johansson, "Event-Based Broadcasting for Multi-Agent Average Consensus," *Automatica*, vol. 49, no. 1, pp. 245–252, 2013.

[21] S. S. Kia, J. Cortés, and S. Martínez, "Distributed Event-Triggered Communication for Dynamic Average Consensus in Networked Systems," *Automatica*, vol. 59, pp. 112–119, 2015.

[22] N. Feamster, J. Rexford, and E. Zegura, "The Road to SDN: A Intellectual History of Programmable Networks," *ACM SIGCOMM Computer Communication Review*, vol. 44, no. 2, pp. 87–98, 2014.

[23] D. Kreutz, F. M. Ramos, P. Verissimo, C. E. Rothenberg, S. Azodolmolky, and S. Uhlig, "Software-Defined Networking: A Comprehensive Survey," *Proceedings of the IEEE*, vol. 103, no. 1, pp. 14–76, 2015.

[24] Open Networking Foundation, "OpenFlow Switch Specification," www.opennetworking.org/wp-content/uploads/2014/10/openflow-spec-v1.3.0.pdf, 2012.

[25] L. Ren, Y. Qin, B. Wang, P. Zhang, P. B. Luh, and R. Jin, "Enabling Resilient Microgrid through Programmable Network," *IEEE Transactions on Smart Grid*, vol. 8, no. 6, pp. 2826–2836, 2017.

[26] Hewlett-Packard, "HP Switch Software OpenFlow Supplement," https://support.hpe.com/hpsc/doc/public/display?docId=emr_na-c03170243, 2012.

[27] J. R. Ballard, I. Rae, and A. Akella, "Extensible and Scalable Network Monitoring Using Opensafe," in *INM/WREN*, 2010.

[28] C. Yu, C. Lumezanu, Y. Zhang, V. Singh, G. Jiang, and H. V. Madhyastha, "Flowsense: Monitoring Network Utilization with Zero Measurement Cost," in *International Conference on Passive and Active Network Measurement*. Springer, pp. 31–41, 2013.

[29] K. Phemius and M. Bouet, "Monitoring Latency with Openflow," in *Proceedings of the 9th International Conference on Network and Service Management (CNSM 2013)*. IEEE, pp. 122–125, 2013.

[30] A. Tootoonchian, M. Ghobadi, and Y. Ganjali, "Opentm: Traffic Matrix Estimator for Openflow Networks," in *International Conference on Passive and Active Network Measurement*. Springer, pp. 201–210, 2010.

[31] Y. Yiakoumis, K.-K. Yap, S. Katti, G. Parulkar, and N. McKeown, "Slicing Home Networks," in *Proceedings of the 2nd ACM SIGCOMM Workshop on Home Networks*. ACM, pp. 1–6, 2011.

[32] K. L. Calvert, W. K. Edwards, N. Feamster, R. E. Grinter, Y. Deng, and X. Zhou, "Instrumenting Home Networks," *SIGCOMM Computer Communication Review*, no. 1, pp. 84–89, 2011.

[33] M. Al-Fares, S. Radhakrishnan, B. Raghavan, N. Huang, and A. Vahdat, "Hedera: Dynamic Flow Scheduling for Data Center Networks." in *NSDI*, vol. 10, no. 2010, 2010, p. 19.

[34] B. Heller, S. Seetharaman, P. Mahadevan, Y. Yiakoumis, P. Sharma, S. Banerjee, and N. McKeown, "Elastictree: Saving Energy in Data Center Networks." *NSDI*, vol. 10, 2010, pp. 249–264.

[35] M. Casado, M. J. Freedman, J. Pettit, J. Luo, N. McKeown, and S. Shenker, "Ethane: Taking Control of the Enterprise," in *ACM SIGCOMM Computer Communication Review*, vol. 37, no. 4. ACM, pp. 1–12, 2007.

[36] R. Sherwood, G. Gibb, K.-K. Yap, et al., "Can the Production Network Be the Testbed?" *OSDI*, vol. 10, 2010, pp. 1–6.

[37] N. Ainsworth and S. Grijalva, "A Structure-Preserving Model and Sufficient Condition for Frequency Synchronization of Lossless Droop Inverter-Based AC Networks," *IEEE Transactions on Power Systems*, vol. 28, no. 4, pp. 4310–4319, 2013.

[38] J. W. Simpson-Porco, F. Dörfler, and F. Bullo, "Synchronization and Power Sharing for Droop-Controlled Inverters in Islanded Microgrids," *Automatica*, vol. 49, no. 9, pp. 2603–2611, 2013.

[39] J. He and Y. W. Li, "An Enhanced Microgrid Load Demand Sharing Strategy," *IEEE Transactions on Power Electronics*, vol. 27, no. 9, pp. 3984–3995, 2012.

[40] H. Mahmood, D. Michaelson, and J. Jiang, "Accurate Reactive Power Sharing in an Islanded Microgrid Using Adaptive Virtual Impedances," *IEEE Transactions on Power Electronics*, vol. 30, no. 3, pp. 1605–1617, 2015.

[41] Y. Li, P. Zhang, and C. Kang, "Compositional Power Flow for Networked Microgrids," *IEEE Power and Energy Technology Systems Journal*, vol. 6, no. 1, pp. 81–84, 2019.

[42] Y. Li, Y. Qin, P. Zhang, and A. Herzberg, "SDN-Enabled Cyber-Physical Security in Networked Microgrids," *IEEE Transactions on Sustainable Energy*, vol. 10, no. 3, pp. 1613–1622, 2019.

[43] C. Ahumada, R. Cárdenas, D. Saez, and J. M. Guerrero, "Secondary Control Strategies for Frequency Restoration in Islanded Microgrids with Consideration of Communication Delays," *IEEE Transactions on Smart Grid*, vol. 7, no. 3, pp. 1430–1441, 2016.

[44] E. A. A. Coelho, D. Wu, J. M. Guerrero, J. C. Vasquez, T. Dragičević, Č. Stefanović, and P. Popovski, "Small-Signal Analysis of the Microgrid Secondary Control Considering a Communication Time Delay," *IEEE Transactions on Industrial Electronics*, vol. 63, no. 10, pp. 6257–6269, 2016.

[45] P.-A. Bliman and G. Ferrari-Trecate, "Average Consensus Problems in Networks of Agents with Delayed Communications," *Automatica*, vol. 44, no. 8, pp. 1985–1995, 2008.

[46] E. Cotilla-Sanchez, P. D. Hines, C. Barrows, S. Blumsack, and M. Patel, "Multi-Attribute Partitioning of Power Networks Based on Electrical Distance," *IEEE Transactions on Power Systems*, vol. 28, no. 4, pp. 4979–4987, 2013.

[47] D. Bian, M. Kuzlu, M. Pipattanasomporn, S. Rahman, and Y. Wu, "Real-Time Co-Simulation Platform Using Opal-RT and Opnet for Analyzing Smart Grid Performance," *Power & Energy Society General Meeting, 2015 IEEE*. IEEE, 2015, pp. 1–5.

[48] M. O. Faruque, T. Strasser, G. Lauss, V. Jalili-Marandi, P. Forsyth, C. Dufour, V. Dinavahi, A. Monti, P. Kotsampopoulos, J. A. Martinez et al., "Real-Time Simulation Technologies for Power Systems Design, Testing, and Analysis," *IEEE Power and Energy Technology Systems Journal*, vol. 2, no. 2, pp. 63–73, 2015.

[49] Hewlett Packard Enterprise Development LP, "Aruba 5400R zl2 Switch Series Data Sheet," www.arubanetworks.com/assets/ds/DS_5400Rzl2SwitchSeries.pdf, 2018.

[50] J. F. Kurose and K. Ross, *Computer Networking: A Top-Down Approach, 7/E*. Pearson, 2017.

5 Formal Analysis of Networked Microgrids Dynamics

The increased penetration of DERs in microgrids and macrogrids leads to new challenges. Power electronic devices are usually used to interface these DERs and some other components such as energy storage systems and programmable loads. Although they enable fast grid control and load changes, the high penetration of power electronic components will reduce the grid inertia significantly, making the utility grid highly sensitive to disturbances and threatening the system stability. These disturbances could be uncontrollable external events, variations in system structure and parameters, disturbances from generation side or consumption, or similar difficulties. Understanding the stability feature of this transforming grid under virtually infinite number of scenarios is a particularly important yet intractable problem.

This chapter elaborates some initial efforts in establishing a tractable method, namely Formal Analysis (FA), for assessing the stability of networked microgrids under uncertainties from heterogeneous sources including DERs [1–3]. In particular, a formal theory with mathematical rigor will be established for computing the bounds of all possible (infinitely many) trajectories and estimating the stability margin for the entire networked microgrid system.

So far, FA has found several application scenarios, including transient stability analysis [3], the identification of stability regions [1], load flow calculation [4], control verification [2], and cybersecurity [5].

5.1 Formal Methods

Among various FA methods, this chapter focuses on reachability analysis, which aims to locate the boundary of a system's possible trajectories when the system is subject to unknown-but-bounded disturbances. Therefore, the goal is to obtain a reachable set as given in Figure 5.1.

When a dynamical system is modeled as a differential-algebraic system of equations (DAEs), one viable solution of calculating its reachable sets may be expressed as follows:

- Without loss of too much generality, the dynamic system could be represented by nonlinear DAEs. Thus, the first step is to abstract the original nonlinear DAEs into linear differential inclusions at each time step.

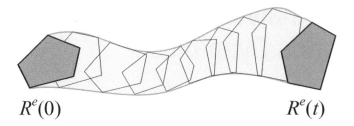

$$R^e(0) \qquad\qquad\qquad\qquad R^e(t)$$

Figure 5.1 Illustration of reachable set.

The purpose of linearization is to obtain $\mathbf{A} = [a_{ij}] \in \mathbb{R}^{n \times n}$, which is the state matrix of the system. The differential inclusion can be obtained through the linearization process:

$$\Delta \dot{\mathbf{x}} \in \mathbf{A} \Delta \mathbf{x} \oplus \mathbf{P}, \tag{5.1}$$

where $\Delta \mathbf{x} = \mathbf{x} - \mathbf{x_0}$, $\mathbf{x_0}$ is the operation point at which the system is linearized; $\mathbf{P}$ is a set of uncertain inputs that can be formulated using a set-based approach that will be introduced later in this chapter; and $\oplus$ is Minkowski addition defined as follows by using two sets E and F:

$$E \oplus F = \{e + f | e \in E, f \in F\}. \tag{5.2}$$

- Based on the abstraction of the system, an analytic form of a reachable set at a time step t_{k+1}, $\mathcal{R}^e(t_{k+1})$, can be derived [6]. In this closed-form formulation, the reachable set $\mathcal{R}^e(t_{k+1})$ is determined by the history term $\mathcal{R}^e(t_k)$ (the reachable set at the previous time step) multiplying $\phi(\mathbf{A}, r)$ (see (5.5)), enlarged by two terms $\Psi(\mathbf{A}, r, \mathbf{p_0})$ and $I_p^e(\mathbf{p_\Delta}, r)$ representing the effects of the deterministic inputs $\mathbf{p_0}$ and the uncertain ones $\mathbf{p_\Delta}$ (see (5.6) and (5.7)), respectively.

The reachable set for a time interval $\tau_k = [t_k, t_{k+1}]$, $\mathcal{R}^e(\tau_k)$, can be obtained by calculating the convex hull of $\mathcal{R}^e(t_k)$ and the solution of the affine dynamics $\phi(\mathbf{A}, r)\mathcal{R}^e(t_k) \oplus \Psi(\mathbf{A}, r, \mathbf{p_0})$ [6], added by an increment I_ξ^e due to the curvature of trajectories for a duration of $r = t_{k+1} - t_k$ (see (5.8)).

In summary, the reachable set at t_{k+1} and that for τ_k can be calculated as follows:

$$\mathcal{R}^e(t_{k+1}) = \phi(\mathbf{A}, r)\mathcal{R}^e(t_k) \oplus \Psi(\mathbf{A}, r, \mathbf{p_0}) \oplus I_p^e(\mathbf{p_\Delta}, r), \tag{5.3}$$

$$\mathcal{R}^e(\tau_k) = C(\mathcal{R}^e(t_k), \phi(\mathbf{A}, r)\mathcal{R}^e(t_k) \oplus \Psi(\mathbf{A}, r, \mathbf{p_0})) \oplus I_p^e(\mathbf{p_\Delta}, r) \oplus I_\xi^e, \tag{5.4}$$

where

$$\phi(\mathbf{A}, r) = e^{\mathbf{A}r}, \tag{5.5}$$

$$\Psi(\mathbf{A}, r, \mathbf{p_0}) = \left\{ \sum_{i=0}^{\eta} \frac{\mathbf{A}^i r^{i+1}}{(i+1)!} \oplus \left[-X(\mathbf{A}, r)r, X(\mathbf{A}, r)r \right] \right\} \mathbf{p_0}, \tag{5.6}$$

$$I_p^e(\mathbf{p}_\Delta, r) = \sum_{i=0}^{\eta} \left(\frac{\mathbf{A}^i r^{i+1}}{(i+1)!} \mathbf{p}_\Delta \right) \oplus \left\{ \left[-X(\mathbf{A}, r)r, X(\mathbf{A}, r)r \right] \cdot \mathbf{p}_\Delta \right\}, \quad (5.7)$$

$$I_\xi^e = \left\{ \left(I \oplus [-X(\mathbf{A}, r), X(\mathbf{A}, r)] \right) \cdot \mathcal{R}^e(t_k) \right\}$$
$$\oplus \left\{ \left(\tilde{I} \oplus [-X(\mathbf{A}, r)r, X(\mathbf{A}, r)r] \right) \cdot \mathbf{p_0} \right\}. \quad (5.8)$$

$$X(\mathbf{A}, r) = e^{|\mathbf{A}|r} - \sum_{i=0}^{\eta} \frac{(|\mathbf{A}|r)^i}{i!}, \quad (5.9)$$

$$I = \sum_{i=2}^{\eta} \left[(i^{\frac{-i}{i-1}} - i^{\frac{-1}{i-1}})r^i, 0 \right] \frac{\mathbf{A}^i}{i!}, \quad (5.10)$$

$$\tilde{I} = \sum_{i=2}^{\eta+1} \left[(i^{\frac{-i}{i-1}} - i^{\frac{-1}{i-1}})r^i, 0 \right] \frac{\mathbf{A}^{i-1}}{i!}. \quad (5.11)$$

Note that $e^{\mathbf{A}r}$ in (5.5) is approximated by using a finite Taylor series $\sum_{i=0}^{\eta} \frac{(\mathbf{A}r)^i}{i!}$ up to order η [6]. The reachable sets calculated through the preceding process are provably overapproximated [7]. Although overapproximation is necessary for formal verification and guarantee of the dynamic system performance, overly conservative results may not be desirable in practice. Reducing the overapproximation of the reachable set while maintaining the efficiency of the algorithm is a problem yet to be fully addressed [8, 9].

5.2 Formal Analysis of Microgrid Dynamics

In this chapter, NMs are assumed to be a system of DAEs where the power-electronic interfaces in the NMs are represented by averaging models. Usually, one can build a system of index-1, semiexplicit, nonlinear DAEs for such NMs, as follows:

$$\begin{cases} \dot{\mathbf{x}} = \mathbf{F}(\mathbf{x}, \mathbf{y}, \mathbf{p}) \\ \mathbf{0} = \mathbf{G}(\mathbf{x}, \mathbf{y}, \mathbf{p}), \end{cases} \quad (5.12)$$

where
$\mathbf{x} \in \mathbb{R}^s$ is the state variable vector (e.g., the integral variable in DER controllers).
$\mathbf{y} \in \mathbb{R}^q$ is the corresponding algebraic variable vector (e.g., the bus voltage).
$\mathbf{p} \in \mathbb{R}^p$ is the corresponding disturbance vector (e.g., PV fluctuations).

The following equations can be obtained by linearizing the NM system at the operation point $(\mathbf{x}^0, \mathbf{y}^0, \mathbf{p}^0)$ by omitting the higher-order Taylor expansions:

$$\begin{cases} \dot{\mathbf{x}} = \mathbf{F}(\mathbf{x}^0, \mathbf{y}^0, \mathbf{p}^0) + \mathbf{F_x} \Delta \mathbf{x} + \mathbf{F_y} \Delta \mathbf{y} + \mathbf{F_p} \Delta \mathbf{p} \\ \mathbf{0} = \mathbf{G}(\mathbf{x}^0, \mathbf{y}^0, \mathbf{p}^0) + \mathbf{G_x} \Delta \mathbf{x} + \mathbf{G_y} \Delta \mathbf{y} + \mathbf{G_p} \Delta \mathbf{p}, \end{cases} \quad (5.13)$$

where the partial derivative matrices $\mathbf{F_x}$, $\mathbf{F_y}$, $\mathbf{F_p}$, $\mathbf{G_x}$, $\mathbf{G_y}$, and $\mathbf{G_p}$ are expressed as follows:

$$\mathbf{F_x} = \frac{\partial \mathbf{F}}{\partial \mathbf{x}}, \qquad \mathbf{F_y} = \frac{\partial \mathbf{F}}{\partial \mathbf{y}}, \qquad \mathbf{F_p} = \frac{\partial \mathbf{F}}{\partial \mathbf{p}},$$

$$\mathbf{G_x} = \frac{\partial \mathbf{G}}{\partial \mathbf{x}}, \qquad \mathbf{G_y} = \frac{\partial \mathbf{G}}{\partial \mathbf{y}}, \qquad \mathbf{G_p} = \frac{\partial \mathbf{G}}{\partial \mathbf{p}}.$$

Since $\mathbf{G_y}^{-1}$ is nonsingular for a system of index-1 DAEs [10], (5.13) can be reformulated by eliminating the second subequation of (5.13):

$$\Delta \dot{\mathbf{x}} = [\mathbf{F_x} - \mathbf{F_y}\mathbf{G_y}^{-1}\mathbf{G_x}]\Delta \mathbf{x} + [\mathbf{F_p} - \mathbf{F_y}\mathbf{G_y}^{-1}\mathbf{G_p}]\Delta \mathbf{p}. \qquad (5.14)$$

Equation (5.14) corresponds to (5.1), where a state matrix $\mathbf{A}_{NMG}$ at each time step has the following form:

$$\mathbf{A}_{NMG} = \mathbf{F_x} - \mathbf{F_y}\mathbf{G_y}^{-1}\mathbf{G_x}. \qquad (5.15)$$

Here $\mathbf{A}_{NMG}$ is equivalent to $\mathbf{A}$ in (5.1) and $[\mathbf{F_p} - \mathbf{F_y}\mathbf{G_y}^{-1}\mathbf{G_p}]\Delta\mathbf{p}$ corresponds to $\mathbf{P}$ in (5.1).

5.2.1   Impact of Disturbances on the State Matrix

The state matrix $\mathbf{A}_{NMG}$ is time varying and has to be updated for each simulation step, making the solution to the reachable set a very expensive process. To improve the computational efficiency, we make use of the fact that the majority of the elements in the state matrix are invariant during the reachability analysis process. At each time step, if we only recalculate those elements of the state matrix that are affected by network disturbances, the burden for updating the state matrix will be largely relieved. Thus, the entire state matrix can be partitioned into two categories of submatrices: those relevant to disturbances and those being always invariant. Without loss of much generality, the submatrices to be updated may consist of those impacted by uncertainties in DER power inputs, loads, and the power exchange of each microgrid at the point of common coupling (PCC), which can be expressed in (5.16):

$$\mathbf{A}_P = \sum_{i=1}^{N_{NMG}} \mathbf{A}_i = \sum_{i=1}^{N_{NMG}} \left(\sum_{j=1}^{N_G} \mathbf{A}_i^{G_j} + \sum_{k=1}^{N_L} \mathbf{A}_i^{L_k} + \mathbf{A}_i^{E} + \mathbf{A}_i^{G,L,E} \right), \qquad (5.16)$$

where
N_{NMG} is the number of microgrids.
N_G is the number of DERs in one microgrid.
N_L is the number of loads in one microgrid.
$\mathbf{A}_i$ is the increment of the state matrix in the ith microgrid.
$\mathbf{A}_i^{G_j}, \mathbf{A}_i^{L_k}, \mathbf{A}_i^{E}$ are the increments only correlated to DERs, loads, and power exchanges at PCC in the ith microgrid.

$\mathbf{A}_i^{G,L,E}$ indicates the cross items, which represent their mutual effects on the matrix's increments.

They can be further expanded as follows:

$$\mathbf{A}_i^{G_j} = \mathbf{F}_{\mathbf{x}}^{G_j} - \mathbf{F}_{\mathbf{y}}^{G_j}\mathbf{G}_{\mathbf{y}}^{-1}\mathbf{G}_{\mathbf{x}}^{G_j} - \mathbf{F}_{\mathbf{y}}^{G_j}\mathbf{G}_{\mathbf{y}}^{-1}\mathbf{G}_{\mathbf{x}}^{C} - \mathbf{F}_{\mathbf{y}}^{C}\mathbf{G}_{\mathbf{y}}^{-1}\mathbf{G}_{\mathbf{x}}^{G_j},$$

$$\mathbf{A}_i^{L_k} = \mathbf{F}_{\mathbf{x}}^{L_k} - \mathbf{F}_{\mathbf{y}}^{L_k}\mathbf{G}_{\mathbf{y}}^{-1}\mathbf{G}_{\mathbf{x}}^{L_k} - \mathbf{F}_{\mathbf{y}}^{L_k}\mathbf{G}_{\mathbf{y}}^{-1}\mathbf{G}_{\mathbf{x}}^{C} - \mathbf{F}_{\mathbf{y}}^{C}\mathbf{G}_{\mathbf{y}}^{-1}\mathbf{G}_{\mathbf{x}}^{L_k},$$

$$\mathbf{A}_i^{E} = \mathbf{F}_{\mathbf{x}}^{E} - \mathbf{F}_{\mathbf{y}}^{E}\mathbf{G}_{\mathbf{y}}^{-1}\mathbf{G}_{\mathbf{x}}^{E} - \mathbf{F}_{\mathbf{y}}^{E}\mathbf{G}_{\mathbf{y}}^{-1}\mathbf{G}_{\mathbf{x}}^{C} - \mathbf{F}_{\mathbf{y}}^{C}\mathbf{G}_{\mathbf{y}}^{-1}\mathbf{G}_{\mathbf{x}}^{E},$$

$$\mathbf{A}_i^{G_j,L_k} = -\mathbf{F}_{\mathbf{y}}^{G_j}\mathbf{G}_{\mathbf{y}}^{-1}\mathbf{G}_{\mathbf{x}}^{L_k} - \mathbf{F}_{\mathbf{y}}^{L_k}\mathbf{G}_{\mathbf{y}}^{-1}\mathbf{G}_{\mathbf{x}}^{G_j},$$

$$\mathbf{A}_i^{G_j,E} = -\mathbf{F}_{\mathbf{y}}^{G_j}\mathbf{G}_{\mathbf{y}}^{-1}\mathbf{G}_{\mathbf{x}}^{E} - \mathbf{F}_{\mathbf{y}}^{E}\mathbf{G}_{\mathbf{y}}^{-1}\mathbf{G}_{\mathbf{x}}^{G_j},$$

$$\mathbf{A}_i^{L_k,E} = -\mathbf{F}_{\mathbf{y}}^{L_k}\mathbf{G}_{\mathbf{y}}^{-1}\mathbf{G}_{\mathbf{x}}^{E} - \mathbf{F}_{\mathbf{y}}^{E}\mathbf{G}_{\mathbf{y}}^{-1}\mathbf{G}_{\mathbf{x}}^{L_k},$$

$$\mathbf{A}_i^{G,L,E} = \mathbf{A}_i^{G_j,L_k} + \mathbf{A}_i^{G_j,E} + \mathbf{A}_i^{L_k,E},$$

where
$\mathbf{F}_{\mathbf{x}}^{G_j}, \mathbf{F}_{\mathbf{y}}^{G_j}, \mathbf{G}_{\mathbf{x}}^{G_j}$ are matrices only related to the uncertainties from the jth DER unit in the ith microgrid.
$\mathbf{F}_{\mathbf{x}}^{L_k}, \mathbf{F}_{\mathbf{y}}^{L_k}, \mathbf{G}_{\mathbf{x}}^{L_k}$ are matrices only related to the changes of the jth load in the ith microgrid.
$\mathbf{F}_{\mathbf{x}}^{E}, \mathbf{F}_{\mathbf{y}}^{E}, \mathbf{G}_{\mathbf{x}}^{E}$ are matrices only related to the disturbances at PCC in the ith microgrid.
$\mathbf{F}_{\mathbf{x}}^{C}, \mathbf{F}_{\mathbf{y}}^{C}, \mathbf{G}_{\mathbf{x}}^{C}$ are constant matrices uncorrelated with any disturbances.

Matrix decomposition as a preprocessing of the state matrix has two major advantages. It is much easier and more efficient to calculate the increment $\mathbf{A}_P$ since usually only a small number of elements need to be updated when disturbances occur. Furthermore, the increment of the state matrix $\mathbf{A}_P$ is decomposed into items correlated to different types of disturbances, and it provides an effective way to quantify the effect of any type of disturbance.

5.2.2   Modeling Disturbances in Networked Microgrids

Properly modeled uncertain inputs are key to FA calculations. For the formal verification purposes, a set-based approach is a suitable choice to handle uncertainties in NMs. Specifically, zonotopes are adopted in the reachable set calculation as they are proven to be computationally stable and efficient, closed under Minkowski operations, and amendable to convex hull computations. Another advantage of using zonotopes is that existing sets for modeling "unknown but bounded" uncertainties in NMs, such as intervals, polytopes, and ellipsoids, can be readily converted to zonotopes.

A zonotope $\mathcal{P}$ is constructed by a center and generators as shown in Figure 5.2 and modeled in (5.17).

$$\mathcal{P} = \left\{ c + \sum_{i=1}^{m} \beta_i g_i \mid \beta_i \in [-1, 1] \right\}, \tag{5.17}$$

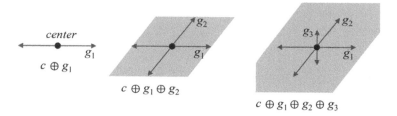

Figure 5.2 Illustration of a zonotope.

where

$c \in \mathbb{R}^n$ is the center.

$g_i \in \mathbb{R}^n$ are the generators.

Therefore, the vector of uncertainties **P** in (5.1) is represented by using a zonotope. If a finer characterization of uncertainties is needed, one can adopt the polynomial zonotype or probabilistic zonotype in the reachable set calculation, which requires modified solution procedures [6].

5.3   Stability Margin Analysis on NMs

An FA calculation will directly produce the reachable sets (or reachtubes) for NM states. For NM operations, especially when the system operates in an islanded mode, it is more important to quantify the distance of an NM system from its stability margin. There are several reasons for this, but we will focus on the following three major ones:

- During any specific period, only if there exists an adequate stability margin can we safely run an NM system.
- If the NM system is found moving toward its stability margin, predictive control or remedial action schemes can be initiated to bring the system back to the stable region.
- Maintaining adequate stability margins is also a necessary condition for NMs to act as resiliency sources for the power grid. Only an NM system with guaranteed stability can proactively provide ancillary services to stabilize, restore, or blackstart the distribution grid to which it is connected.

5.3.1   Quasi diagonalized Geršgorin Theorem

In order to assess the stability margin, an improved Geršgorin technique is devised to provide a quick estimation of the eigenvalues of a disturbed NM system.

At a particular time step, the eigenvalues of an NM system can be calculated through the following standard procedures for analyzing the small signal stability of a power system [11]:

$$\begin{cases} \mathbf{A}\mathbf{v}_i = \lambda_i \mathbf{v}_i \\ \mathbf{A}^{\mathrm{T}}\mathbf{u}_i = \lambda_i \mathbf{u}_i, \end{cases} \tag{5.18}$$

where

λ_i is the ith generalized eigenvalue of the NM system.

$\mathbf{v}_i$ and $\mathbf{u}_i^T$ are the ith right and left eigenvector, respectively.

The right and left eigenvectors satisfy the orthogonal normalization conditions:

$$\begin{cases} \mathbf{u}_i^T \mathbf{v}_j = \delta_{ij} \\ \mathbf{u}_i^T \mathbf{A} \mathbf{v}_j = \delta_{ij} \lambda_i, \end{cases} \tag{5.19}$$

where

δ_{ij} is the Kronecker sign.

An exact eigenanalysis is computationally expensive. For example, the overall complexity of the QR algorithm, a widely adopted algorithm to determine the eigenvalue of a complex matrix, is of $\mathcal{O}(n^3)$. The calculation of the exact eigenvalues sometimes is unnecessary, especially if it is known a priori that the NM system is far away from its stability margin. So, for an NM system with state matrix $\mathbf{A}$, instead of calculating the exact eigenvalues, we only need to estimate the ranges of the eigenvalues using the Geršgorin disks and the Geršgorin set, as follows.

Geršgorin's Theorem [12]: *Consider a nonsingular square matrix $\mathbf{A} \in \mathbb{C}^{n \times n}$, define $r_k(\mathbf{A}) \doteq \sum_{j \in N \setminus \{k\}} |a_{kj}|$, and define the kth row Geršgorin disk $\Gamma_k(\mathbf{A}) \doteq \{|x - a_{kk}| \leq r_k(\mathbf{A}), x \in \mathbb{R}\}$. If λ_i is the ith eigenvalue of $\mathbf{A}$, then there exists $k \in N = \{1, 2, \dots, n\}$ such that*

$$|\lambda_i - a_{kk}| \leq r_k(\mathbf{A}); \tag{5.20}$$

and the spectrum $\sigma(\mathbf{A})$ of $\mathbf{A}$ lies in the Geršgorin set of $\mathbf{A}$:

$$\sigma(\mathbf{A}) \subseteq \Gamma(\mathbf{A}) \doteq \bigcup_{k-1}^{n} \Gamma_k(\mathbf{A}). \tag{5.21}$$

Here the Geršgorin set $\Gamma(\mathbf{A})$ is the union of all the Geršgorin disks.

Figure 5.3 is given to illustrate the Geršgorin disks, where "Exact eigenvalue" shows the actual eigenvalues of a system before disturbances, and the "Geršgorin disk" shows the possible locations of each system's eigenvalues under disturbances.

Figure 5.3 demonstrates that the estimation of eigenvalue distribution can be overly conservative because the state matrix is usually neither diagonal nor strictly diagonal

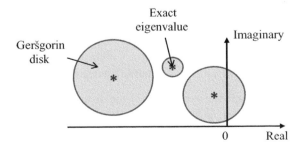

Figure 5.3 Illustration of Geršgorin sets.

dominant. This means that we need to reduce the cooperativeness of the the Geršgorin set so that the estimation of eigenvalue locations becomes more accurate, which can significantly improve the efficiency of the stability margin estimation. To this end, we have developed a quasi diagonalized Geršgorin method to significantly tighten the Geršgorin disks.

The state matrix $\mathbf{A}$ of a disturbed system can be decomposed into $\mathbf{A}_0$, which is the state matrix of the undisturbed system operating at $(\mathbf{x}_0, \mathbf{y}_0)$, and $\mathbf{A}_P$ denoting the increment of the state matrix constructed based on a bounded set of uncertainties. Assume $\mathbf{S}_0$, $\mathbf{U}_0^T$, and $\mathbf{V}_0$ are the eigenvalue matrix, left eigenvector matrix, and right eigenvector matrix at $(\mathbf{x}_0, \mathbf{y}_0)$, respectively. $\mathbf{A}$ can then be quasi diagonalized by left-multiplying $\mathbf{U}_0^T$ and right-multiplying $\mathbf{V}_0$. Bearing in mind the orthogonal normalization conditions in (5.19), we can obtain the following:

$$\mathbf{U}_0^T \mathbf{A} \mathbf{V}_0 = \mathbf{U}_0^T \mathbf{A}_0 \mathbf{V}_0 + \mathbf{U}_0^T \mathbf{A}_P \mathbf{V}_0 = \mathbf{S}_0 + \mathbf{S}_P, \tag{5.22}$$

where $\mathbf{S}_P$ is the increment of the eigenvalue matrix.

Through the preceding quasi diagonalization process, the Geršgorin estimation of a disturbed NM system is converted to the eigenanalysis of $\mathbf{S}_P$, the increment of the eigenvalue matrix due to uncertainties. The Geršgorin disks and set of $\mathbf{S}_P$ can then be evaluated by the following:

$$\Gamma_k(\mathbf{S}_P) = \{|x - s_{kk}| \le r_k(\mathbf{S}_P), x \in \mathbb{R}\}, \tag{5.23}$$

$$\sigma_k(\mathbf{S}_P) \subseteq \Gamma(\mathbf{S}_P) \doteq \bigcup_{k=1}^{n} \Gamma_k(\mathbf{S}_P). \tag{5.24}$$

In summary, with the quasi diagonalization, the distribution of each eigenvalue in a system under uncertainties can be expressed as a Geršgorin disk with $\mathbf{S}_0$ as its center and $\Gamma_k(\mathbf{S}_P)$ as its corresponding area. This will normally lead to less conservative Geršgorin disks.

5.3.2   Stability Margin Calculation

Since the quasi diagonalized Geršgorin algorithm can effectively estimate the eigenvalues of an NM system, it can be used to decide whether the boundary of the reachable set at a specific time remains in the stable area. Therefore, a practical approach to probing the stability margins for an NM system can be developed by combining the formal analysis with the quasi diagonalized Geršgorin algorithm. The following steps will be performed in the process of finding the stability margin:

- Perform FA to obtain the reachable set $\mathcal{R}^e(t_k)$ for the disturbed NMs.
- Extract the vertices of the reachable set, which will be used as inputs for a quasi diagonalized Geršgorin calculation in (5.23) and (5.24).
- Evaluate the Geršgorin disks and set to determine if the system is stable under a specific level of disturbances.

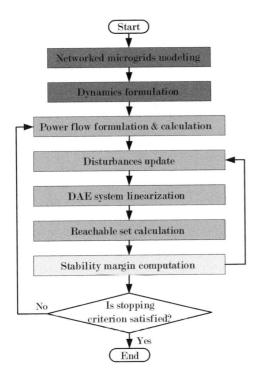

Figure 5.4 The procedures of stability margin assessment based on reachability analysis.

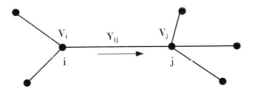

Figure 5.5 Illustration of a power flow formulation.

Figure 5.4 summarizes the analysis process of the stability margin through the FA enhanced by the quasi diagonalized Geršgorin algorithm.

As shown in Figure 5.4, the algebraic equation of an NM system is first established to model NM elements such as feeder sections, transformers, and loads; then, a set of ordinary differential equations are built to formulate the dynamics of distributed generators, components and the associated inverters, converters, and various controllers. Next, an extended-admittance-matrix-based power flow algorithm is performed to obtain initial conditions of the NM system.

Power Flow Formulation and Calculation

Consider the branch between node i and node j in Figure 5.5. The branch admittance is $Y_{ij} = |Y_{ij}| \cos(\alpha_{ij}) + j|Y_{ij}| \sin(\alpha_{ij})$; the voltage of node i is expressed as $V_i = |V_i| \cos(\theta_i) + j|V_i| \sin(\theta_i)$; and the voltage of node j is expressed as

$V_j = |V_j| \cos(\theta_j) + j|V_j| \sin(\theta_j)$. Then, the current injection of each node can be expressed as follows:

$$\begin{bmatrix} I_i \\ I_j \end{bmatrix} = \begin{bmatrix} -Y_{ij} & Y_{ij} \\ Y_{ij} & -Y_{ij} \end{bmatrix} \begin{bmatrix} V_i \\ V_j \end{bmatrix}. \tag{5.25}$$

Correspondingly, the power flow through branch *i-j* can be calculated via the following equation:

$$P_j + jQ_j = V_j \bar{I}_j = V_j \overline{\{Y_{ij}(V_i - V_j)\}} = V_j \bar{V}_i \bar{Y}_{ij} - V_j \bar{V}_j \bar{Y}_{ij}. \tag{5.26}$$

As an example, the first item $V_j \bar{V}_i \bar{Y}_{ij}$ is expanded:

$$V_j \bar{V}_i \bar{Y}_{ij} = V_i V_j |Y_{ij}| \cos(\theta_i - \theta_j - \alpha_{ij}) + j V_i V_j |Y_{ij}| \sin(\theta_i - \theta_j - \alpha_{ij}). \tag{5.27}$$

We can then obtain the power flow equation in (5.28):

$$\begin{bmatrix} |Y_{ij}| \cos(\theta_i - \theta_j - \alpha_{ij}) \\ |Y_{ij}| \sin(\theta_i - \theta_j - \alpha_{ij}) \end{bmatrix} \cdot \begin{bmatrix} V_{ij} \end{bmatrix} \circ \begin{bmatrix} V_{ij} \\ V_{ij} \end{bmatrix} + \begin{bmatrix} P_{ij}^G \\ Q_{ij}^G \end{bmatrix} - \begin{bmatrix} P_{ij}^L \\ Q_{ij}^L \end{bmatrix}$$

$$= \overline{\mathbf{Y}} \cdot \mathbf{V} \circ \overline{\mathbf{V}} + \overline{\mathbf{S}^G} - \overline{\mathbf{S}^L} = \mathbf{0}, \tag{5.28}$$

Here $\circ$ denotes the Hadamard product as follows:

$$((E \circ F)_{ij} = [e_{ij} \cdot f_{ij}], e_{ij} \in E, f_{ij} \in F). \tag{5.29}$$

P_{ij}^G and Q_{ij}^G are the active and reactive power injections from node *i* to node *j*, respectively.
P_{ij}^L and Q_{ij}^L are the active and reactive power loads at node *j*, respectively, and $\overline{\mathbf{Y}}$ is the extended admittance matrix and can be expressed in (5.30).

$$\overline{\mathbf{Y}} = \begin{bmatrix} |Y_{11}| \cos \beta_{11} & |Y_{12}| \cos \beta_{12} & \cdots & |Y_{1n}| \cos \beta_{1n} \\ |Y_{21}| \cos \beta_{21} & |Y_{22}| \cos \beta_{22} & \cdots & |Y_{2n}| \cos \beta_{2n} \\ \vdots & \vdots & \ddots & \vdots \\ |Y_{n1}| \cos \beta_{n1} & |Y_{n2}| \cos \beta_{n2} & \cdots & |Y_{nn}| \cos \beta_{nn} \\ |Y_{11}| \sin \beta_{11} & |Y_{12}| \sin \beta_{12} & \cdots & |Y_{1n}| \sin \beta_{1n} \\ |Y_{21}| \sin \beta_{21} & |Y_{22}| \sin \beta_{22} & \cdots & |Y_{2n}| \sin \beta_{2n} \\ \vdots & \vdots & \ddots & \vdots \\ |Y_{n1}| \sin \beta_{n1} & |Y_{n2}| \sin \beta_{n2} & \cdots & |Y_{nn}| \sin \beta_{nn} \end{bmatrix}, \tag{5.30}$$

where $\beta_{ij} = \theta_i - \theta_j - \alpha_{ij}$.

From (5.30), it can be seen that, if a subsystem has *n* nodes, the dimensions of its corresponding matrix $\mathbf{Y}$ will be $2n \times n$.

To better explain (5.30), the following three-bus system is given as an example to show how to get matrix $\overline{\mathbf{Y}}$.

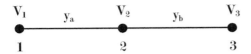

Figure 5.6 Three-bus system.

Example 6.1:

If we assume the admittances shown in Figure 5.6 are y_a and y_b, the corresponding admittance matrix can be expressed as follows:

$$\begin{bmatrix} Y_{11} & Y_{12} & Y_{13} \\ Y_{21} & Y_{22} & Y_{23} \\ Y_{31} & Y_{32} & Y_{33} \end{bmatrix} = \begin{bmatrix} -y_a & y_a & 0 \\ y_a & -(y_a + y_b) & y_b \\ 0 & y_b & -y_b \end{bmatrix}, \tag{5.31}$$

where each element can be further expressed as follows:

$$Y_{11} = |Y_{11}|\cos(\alpha_{11}) + j|Y_{11}|\sin(\alpha_{11}) = -y_a,$$

$$Y_{12} = |Y_{12}|\cos(\alpha_{12}) + j|Y_{12}|\sin(\alpha_{12}) = y_a,$$

$$Y_{13} = |Y_{13}|\cos(\alpha_{13}) + j|Y_{13}|\sin(\alpha_{13}) = 0,$$

$$Y_{21} = |Y_{21}|\cos(\alpha_{21}) + j|Y_{21}|\sin(\alpha_{21}) = y_a,$$

$$Y_{22} = |Y_{22}|\cos(\alpha_{22}) + j|Y_{22}|\sin(\alpha_{22}) = -(y_a + y_b),$$

$$Y_{23} = |Y_{23}|\cos(\alpha_{23}) + j|Y_{23}|\sin(\alpha_{23}) = y_b,$$

$$Y_{31} = |Y_{31}|\cos(\alpha_{31}) + j|Y_{31}|\sin(\alpha_{31}) = 0,$$

$$Y_{32} = |Y_{32}|\cos(\alpha_{32}) + j|Y_{32}|\sin(\alpha_{32}) = y_b,$$

$$Y_{33} = |Y_{33}|\cos(\alpha_{33}) + j|Y_{33}|\sin(\alpha_{33}) = -y_b,$$

Then, based on the preceding equations, $|Y_{ij}|$ and α_{ij} can be obtained. So, the matrix $\overline{\mathbf{Y}}$ given in (5.30) can be expressed as follows:

$$\begin{bmatrix} |Y_{11}|\cos(-\alpha_{11}) & |Y_{12}|\cos(\theta_1 - \theta_2 - \alpha_{12}) & 0 \\ |Y_{21}|\cos(\theta_2 - \theta_1 - \alpha_{21}) & |Y_{22}|\cos(-\alpha_{22}) & |Y_{23}|\cos(\theta_2 - \theta_3 - \alpha_{23}) \\ 0 & |Y_{32}|\cos(\theta_3 - \theta_2 - \alpha_{32}) & |Y_{33}|\cos(-\alpha_{33}) \\ |Y_{11}|\sin(-\alpha_{11}) & |Y_{12}|\sin(\theta_1 - \theta_2 - \alpha_{12}) & 0 \\ |Y_{21}|\sin(\theta_2 - \theta_1 - \alpha_{21}) & |Y_{22}|\sin(-\alpha_{22}) & |Y_{23}|\sin(\theta_2 - \theta_3 - \alpha_{23}) \\ 0 & |Y_{32}|\sin(\theta_3 - \theta_2 - \alpha_{32}) & |Y_{33}|\sin(-\alpha_{33}) \end{bmatrix}. \tag{5.32}$$

The extended-admittance-matrix-based power flow formulation has the following advantages:

- The formulation of admittance is modular and thus particularly suitable for the online modeling of NM systems where "plug-and-play" and the removal of NM components (e.g., DERs and even individual microgrids) are common and frequent.

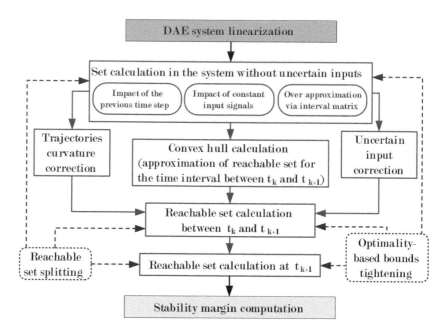

Figure 5.7 Reachability analysis of a linearized NM system.

- The formulation allows the direct assessment of power flow under uncertainties. If P_{ij}^G and Q_{ij}^G are zonotope inputs, the power flow zonotopes that enclose the effects of disturbances can be directly obtained through (5.28).

More details of the extended-admittance-matrix-based power flow calculation can be found in [13, 14].

Reachable Set Calculation

Once the power flow results are obtained, the reachable set calculation for the NM system can be performed. For a linearized NM system, which has a form shown in (5.13), its reachable set can be obtained by using (5.3) and (5.4). The corresponding calculation flowchart can be found in Figure 5.7, with details given in (5.5)–(5.11).

Stability Margin Computation

Once the reachable set at t_{k+1} is calculated, the eigenvalue distribution at the edge of the reachable set can then be estimated by using the quasi diagonalization-based Geršgorin approach. The probing of stability margin is shown in Figure 5.8 and described as follows:

- Check whether the NM system satisfies the stability criterion (i): First, identify s_{kk}^{max}, which is the real part of the center of the rightmost Geršgorin disk. Then, shift s_{kk}^{max} by the radius of the rightmost Geršgorin disk $r_k^{max}(\mathbf{S}_P)$, checking whether the largest real part of the possible eigenvalues is less than a given threshold α_0:

$$s_{kk}^{max} + r_k^{max}(\mathbf{S}_P) \leq \alpha_0 \quad \text{(Stability criterion (i))}, \tag{5.33}$$

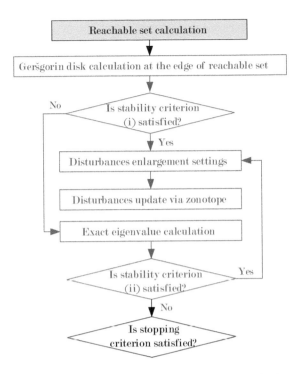

Figure 5.8 Process of establishing the stability margin through an augmented eigenanalysis with FA.

If (i) is true, then the system must be stable. If (i) is false, then the system may or may not be stable, meaning a detailed QR analysis is needed to further verify the system stability with the exact eigenvalue information.

- If the NM system is surely stable, we can enlarge disturbances to further probe the stability margin. Given the increased level of disturbances, conduct the reachability analysis and the Geršgorin estimation again and check whether the stability criterion (i) is satisfied.

- If the stability criterion (i) is not satisfied, perform the exact eigenanalysis and check whether the stability criterion (ii) is satisfied:

$$\alpha_{\max} \leq \alpha_0 \quad \text{(Stability criterion (}ii\text{)),} \tag{5.34}$$

where $\alpha_{\max}$ is the largest (least negative) real part of the eigenvalues.

The aforementioned calculation will continue until it reaches the end time or the NM system is found unstable after certain time steps. If one of the conditions is met, then terminate the process; otherwise, the power flow and reachable set calculations are repeated.

The efficiency of the stability margin analysis of a disturbed NM system is improved by using the quasi diagonalization-based Geršgorin theory because of the following reasons:

- If an exact eigenanalysis is adopted, whenever a disturbance occurs, the full process of eigenvalue computations including the update of the entire state matrix,

Householder transformation, Hessenberg matrix formation, QR decomposition, and so on [15] should be performed to compute the exact eigenvalues. Differently, when the quasi diagonalization-based Geršgorin algorithm is used, only the increment of the state matrix needs to be calculated. Thus, eigenvalues can be efficiently estimated, making it unnecessary to undergo the complex procedures associated with calculating exact eigenvalues.

- The exact eigenvalue calculation is not always required. During the probing process to identify the stability margin, oftentimes a quick quasi diagonalized Geršgorin estimation would indicate that the largest real part of the eigenvalues resides in the left half plane and far away from the y-axis. Because the Geršgorin set provably covers all possible eigenvalues, the system would be absolutely stable. The exact eigenanalysis is therefore unnecessary for most of the time and it is only needed when the NM system is pushed sufficiently close to the stability boundary.

5.4   Distributed Formal Analysis (DFA)

Although FA is a potent tool which potentially outperforms energy-based methods [16] and time-domain simulations [17], a centralized FA could still be too expensive to directly evaluate the stability of a large-scale or configurable power system with massive DERs. Moreover, securing data and information of a microgrid has become a priority for today's microgrid stakeholders. The need to protect the data privacy of microgrids, however, makes a centralized stability assessment impractical. Efficiently analyzing the stability of a large-scale NM system while ensuring the privacy of information involved in calculation is an urgent problem.

Distributed FA (or compositional FA) has originally been developed in the FA community [3, 6]. For instance, [6] has discussed two compositional FA techniques. One technique focuses on the compositional computation the set of linearization errors because the linearization error computation has a complexity of $\mathcal{O}(n^5)$. In this approach, computing of the linear differential inclusions is based on the full model, as the reachability analysis excluding the calculation of the linearization error has a complexity of $\mathcal{O}(n^3)$ [6]. The other technique is to partition the interconnected power systems into subsystems so that the reachable sets for the subsystems are computed in a distributed way [3]. In this book, based on the centralized FA, we introduce a distributed FA (DFA) method by decoupling a large NM system into subsystems and calculating their reachable sets.

If a large NM system is partitioned into $N+M$ smaller subsystems, the reachable sets of the overall NMs ($\mathcal{R}_s^e(t_{k+1})$ at each time step and $\mathcal{R}_s^e(\tau_k)$ during time steps) can be composed by using the Cartesian product of the reachable sets of each subsystem ($\mathcal{R}_i^e(t_{k+1})$ at each time step and $\mathcal{R}_i^e(\tau_k)$ during time steps) as expressed in (5.35) and (5.36):

$$\mathcal{R}_s^e(t_{k+1}) = \varphi_1 \mathcal{R}_1^e(t_{k+1}) \times \varphi_2 \mathcal{R}_2^e(t_{k+1}) \times \cdots \times \varphi_{N+M} \mathcal{R}_{N+M}^e(t_{k+1}), \qquad (5.35)$$

$$\mathcal{R}_s^e(\tau_k) = \varphi_1 \mathcal{R}_1^e(\tau_k) \times \varphi_2 \mathcal{R}_2^e(\tau_k) \times \cdots \times \varphi_{N+M} \mathcal{R}_{N+M}^e(\tau_k). \qquad (5.36)$$

Here a relationship matrix φ_i, which contains 1s when states are correlated and 0s when they are not, maps the states of the ith subsystem to those of the overall NM system.

After the reachable sets are calculated, a distributed enhanced quasi diagonalized Geršgorin (DQG) theory is also devised accordingly to help evaluate the stability margin via DFA. Assume again that the large interconnected system is partitioned into $N+M$ smaller subsystems. The Geršgorin set of the overall NM system $\Gamma(\mathbf{S_P})$ can then be established through the Cartesian product ($\times$) of the Geršgorin set $\Gamma(\mathbf{S_{i,P}})$ of each subsystem, as shown in (5.37).

$$\Gamma(\mathbf{S_P}) = \varphi_1\Gamma(\mathbf{S_{1,P}}) \times \varphi_2\Gamma(\mathbf{S_{2,P}}) \times \cdots \times \varphi_{N+M}\Gamma(\mathbf{S_{N+M,P}}), \qquad (5.37)$$

Calculation details of Geršgorin disk and sets in each subsystem can be found in (5.23) and (5.24).

5.5   Partitioning a Large Networked Microgrids System

Performing a reachability analysis of a large NM system can be burdensome. To resolve the problem, the grid decomposition is introduced to develop a scalable DFA that can enable a faster stability evaluation of NMs. Many decomposition techniques were presented to partition the original network for distributed calculation, such as the coherency-based decomposition algorithm [18], the hierarchical spectral clustering approach [19], the Multi-Area Thévenin Equivalence (MATE) method [20], the waveform relaxation technique [21], and so on. Here, an $N+M$ decomposition is introduced as an example.

5.5.1 $N+M$ Decomposition

An $N+M$ decomposition is presented to divide a large NM system into several subsystems to overcome the limitations of existing techniques. Subsystems are coupled according to each subsystem's power injection, as shown in Figure 5.9.

In an NM system, each microgrid consists of at least one DER and microgrids are networked through backbone feeders. Naturally, the microgrids in the NM system can be divided into N active subsystems where an active subsystem could be a microgrid or a part of a microgrid consisting of at least one DER. The backbone network of the NM system excluding all DERs can be partitioned into M passive subsystems. This is the so-called $N+M$ decomposition, as shown in Figure 5.10. When the NM system is not large, the $N+M$ decomposition can be quite straightforward. However, if the system is large, the balance of computational burden for each subsystem might be considered. One may analyze the computational complexity for simulating subsystems and the communication costs, and an optimal $N+M$ decomposition can be obtained by solving a graph partitioning problem [20].

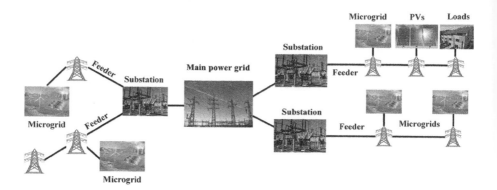

Figure 5.9 Original power systems integrated with NMs.

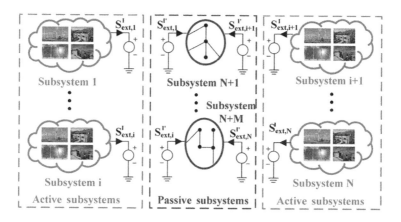

Figure 5.10 Decoupled subsystems using $N+M$ decomposition.

Once the NM system is divided into $N+M$ subsystems, the algebraic (power flow) equations of the overall NM system can be formulated as follows:

$$\mathbf{Y_{ext}} \cdot \mathbf{V_{ext}} \circ \overline{\mathbf{V}}_{\mathbf{ext}} + \mathbf{S^G_{ext}} - \mathbf{S^L_{ext}} - \mathbf{S^I_{ext}} = \mathbf{0}, \qquad (5.38)$$

where $\circ$ is the Hadamard product as shown in (5.29).

Compared with (5.28), the major difference between (5.38) and (5.28) is the introduction of the item $\mathbf{S^I_{ext}}$, which is caused by system decomposition. In the following, the variables in (5.38) are introduced in detail.

Extended Admittance Matrix

After the $N+M$ decomposition, the extended admittance matrix $\mathbf{Y_{ext}}$ in (5.38) becomes a block diagonal matrix. The first N blocks $\mathbf{Y_{11}}, \ldots, \mathbf{Y_{NN}}$ correspond to those extended admittance matrices of the N active subsystems, while the following

M blocks $Y_{N+1,N+1}, \ldots, Y_{N+M,N+M}$ are those of the M passive subsystems. The matrix form is shown as follows:

$$
Y_{ext} = \begin{bmatrix}
Y_{11} & \cdots & 0 & 0 & \cdots & 0 \\
\vdots & \ddots & \vdots & \vdots & \ddots & \vdots \\
0 & \cdots & Y_{NN} & 0 & \cdots & 0 \\
0 & \cdots & 0 & Y_{N+1,N+1} & \cdots & 0 \\
\vdots & \ddots & \vdots & \vdots & \ddots & \vdots \\
0 & \cdots & 0 & 0 & \cdots & Y_{N+M,N+M}
\end{bmatrix},
\tag{5.39}
$$

Extended Voltage Vectors

Assume the bus voltage vectors of the active subsystems are $V_{11}, \ldots, V_{NN}$, and the voltage vectors of the passive subsystems are $V_{N+1,N+1}, \ldots, V_{N+M,N+M}$. Then the bus voltage vector $\mathbf{V}_{ext}$ and the extended bus voltage vector $\overline{\mathbf{V}}_{ext}$ of the NM system can be expressed as follows:

$$
\mathbf{V}_{ext} = \begin{bmatrix} V_{11}, & \cdots, & V_{NN}, & V_{N+1,N+1}, & \cdots, & V_{N+M,N+M} \end{bmatrix}^T,
\tag{5.40}
$$

$$
\overline{\mathbf{V}}_{ext} = \begin{bmatrix} V_{11}, & V_{11}, & \cdots, & V_{NN}, & V_{NN}, \\ V_{N+1,N+1}, & V_{N+1,N+1}, & \cdots, & V_{N+M,N+M}, & V_{N+M,N+M} \end{bmatrix}^T,
\tag{5.41}
$$

Extended Power Vectors

$\mathbf{S}_{ext}^{G}$ is the vector containing the power injections from DERs to active subsystems. $\mathbf{S}_{ext}^{L}$ in (5.38) denotes the load power vector. $\mathbf{S}_{ext}^{I}$ is the vector of power exchange at the interfaces of subsystems. Please note that the values of $\mathbf{S}_{ext,i}^{I}$ and those of $\mathbf{S}_{ext,i}^{I'}$ (see Figure 5.10) are normally close and are identical if we ignore the losses across tie lines. The entries in $\mathbf{S}_{ext}^{I}$ are more or less interdependent with each other and therefore the coupling effect should be considered at each step of the formal analysis process. To make sure those entries are updated together, a status flag technique will be introduced as detailed in the next section.

5.5.2 Partitioning a Large NM System

The $N+M$ decomposition gives M passive subsystems without DERs, meaning those entries in $\mathbf{S}_{ext}^{G}$ corresponding to the nodes in the M subsystems are all zeros. Therefore, we can formulate the algebraic equations of the active and passive subsystems separately, as follows:

$$
\begin{cases}
\mathbf{Y}_{kk} \cdot \mathbf{V}_{kk} \circ \overline{\mathbf{V}}_{kk} + \mathbf{S}_{kk}^{G} - \mathbf{S}_{kk}^{L} - \mathbf{S}_{kk}^{I} = 0 \\
\qquad \mathbf{Y}_{jj} \cdot \mathbf{V}_{jj} \circ \overline{\mathbf{V}}_{jj} - \mathbf{S}_{jj}^{L} - \mathbf{S}_{jj}^{I} = 0, \\
\mathbf{k} = 1, \ldots, N, \mathbf{j} = N + 1, \ldots, N + M.
\end{cases}
\tag{5.42}
$$

As the power flows at the subsystem interfaces are introduced to link subsystems, all the subsystems are well decoupled so that the admittance blocks of subsystems ($\mathbf{Y_{kk}}$ and $\mathbf{Y_{jj}}$) are fully independent with each other (see (5.42)). Therefore, the essential calculation is to obtain and update $\mathbf{S}_{\mathbf{ext}}^{\mathbf{I}}$.

5.5.3   Modeling of Each Subsystem

For a partitioned NM system, if the averaging models [22] for DER converters are adopted, the ith subsystem can still be represented as a semiexplicit, index-1 DAE system. Let $\mathbf{x_i} \in \mathbb{R}^{s_i}$ be the state variables (e.g., integral variable in DER controllers), $\mathbf{y_i} \in \mathbb{R}^{q_i}$ be the algebraic variables (e.g., bus voltages), and $\mathbf{p_i} \in \mathbb{R}^{p_i}$ be the disturbances (e.g., PV power variations). Then the DAE description of the ith subsystem is as follows:

$$\begin{cases} \dot{x}_i = \mathbf{f_i}(\mathbf{x_i}, \mathbf{y_i}, \mathbf{p_i}) \\ 0 = \mathbf{g_i}(\mathbf{x_i}, \mathbf{y_i}, \mathbf{p_i}). \end{cases} \tag{5.43}$$

The DAE subsystem is then linearized around the operation point $(\mathbf{x_i^0}, \mathbf{y_i^0}, \mathbf{p_i^0})$ at each step [6] to facilitate the reachability analysis. Define $\mathbf{f_{x_i}}$, $\mathbf{f_{y_i}}$, $\mathbf{f_{p_i}}$, $\mathbf{g_{x_i}}$, $\mathbf{g_{y_i}}$, and $\mathbf{g_{p_i}}$ as the matrices of partial derivatives with respect to different variables, that is, $\mathbf{f_{x_i}} = \partial \mathbf{f_i}/\partial \mathbf{x_i}$, $\mathbf{f_{y_i}} = \partial \mathbf{f_i}/\partial \mathbf{y_i}$, $\mathbf{f_{p_i}} = \partial \mathbf{f_i}/\partial \mathbf{p_i}$, $\mathbf{g_{x_i}} = \partial \mathbf{g_i}/\partial \mathbf{x_i}$, $\mathbf{g_{y_i}} = \partial \mathbf{g_i}/\partial \mathbf{y_i}$, and $\mathbf{g_{p_i}} = \partial \mathbf{g_i}/\partial \mathbf{p_i}$. Then the linearized subsystem equations are as follows:

$$\begin{cases} \dot{x}_i = \mathbf{f_i}(\mathbf{x_i^0}, \mathbf{y_i^0}, \mathbf{p_i^0}) + \mathbf{f_{x_i}} \Delta \mathbf{x_i} + \mathbf{f_{y_i}} \Delta \mathbf{y_i} + \mathbf{f_{p_i}} \Delta \mathbf{p_i} \\ 0 = \mathbf{g_i}(\mathbf{x_i^0}, \mathbf{y_i^0}, \mathbf{p_i^0}) + \mathbf{g_{x_i}} \Delta \mathbf{x_i} + \mathbf{g_{y_i}} \Delta \mathbf{y_i} + \mathbf{g_{p_i}} \Delta \mathbf{p_i}. \end{cases} \tag{5.44}$$

An index-1 DAE system in (5.43) has an invertible $\mathbf{g_{y_i}}$ [10]. Thus (5.44) can be rewritten as follows [6]:

$$\Delta \dot{x}_i = [\mathbf{f_{x_i}} - \mathbf{f_{y_i}} \mathbf{g_{y_i}^{-1}} \mathbf{g_{x_i}}] \Delta \mathbf{x_i} + [\mathbf{f_{p_i}} - \mathbf{f_{y_i}} \mathbf{g_{y_i}^{-1}} \mathbf{g_{p_i}}] \Delta \mathbf{p_i}. \tag{5.45}$$

It can be seen that the state matrix of the ith subsystem is $\mathbf{A_i} = \mathbf{f_{x_i}} - \mathbf{f_{y_i}} \mathbf{g_{y_i}^{-1}} \mathbf{g_{x_i}} = [a_{jk}] \in \mathbb{R}^{q_i \times q_i}$ and the uncertain inputs of the subsystem is defined as $\mathbf{P_i} = [\mathbf{f_{p_i}} - \mathbf{f_{y_i}} \mathbf{g_{y_i}^{-1}} \mathbf{g_{p_i}}] \Delta \mathbf{p_i}$.

If the uncertain inputs are modeled by sets, then we can construct a differential inclusion by using (5.45) to abstract the ith subsystem subject to disturbances, as follows:

$$\Delta \dot{x}_i \in \mathbf{A_i} \Delta \mathbf{x_i} \oplus \mathbf{P_i}. \tag{5.46}$$

Most recently, some literature started to investigate the use of a system of ordinary differential equations (ODEs) for more accurate modeling of microgrids and networked microgrids [23], which might be a promising direction. More results in this direction can be included in the second edition of this book.

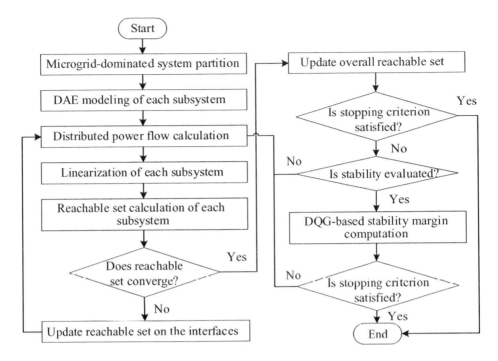

Figure 5.11 Flowchart of DFA calculation.

5.6   Implementation of DFA for Networked Microgrids Analysis

5.6.1   Procedure of Calculation

Figure 5.11 demonstrates the process of the DQG-enabled DFA. Initially, the $N+M$ decomposition is performed to divide a networked microgrid system into multiple subsystems. The power flows in all the subsystems are then calculated in tandem taking into consideration the data interchange among subsystems. Next, the DAEs of subsystems are linearized through (5.44). Subsequently, the DQG-enabled DFA is implemented, including three substeps:

- Perform DFA to compute the set of reachable states of each subsystem.
  If the state sets at the interfaces converge, the state sets of the NM system can then be computed by (5.35) and (5.36). If not, the algebraic equations are recalculated, and the power flow results are passed to the DFA module to update the state sets in each subsystem.
- Sequentially, Geršgorin disks are obtained by conducting DQG at the vertices of reachable sets.
- Finally, the stability of the disturbed NM system is checked by using the Geršgorin disks.

5.6.2   Distributed Algorithm and Data Exchange in DFA

A double-loop iterative process is adopted by the distributed formal analysis. The outer iteration is responsible for updating the power exchanges at interfaces and check whether convergence is achieved. The inner loop calculates power flows of the passive network and obtains the state sets of subsystems.

The following explains how the convergence is checked in the outer loop. Assume $\mathbf{V_{ki}} = [\mathbf{V_k}, \mathbf{V_i}]$ is the voltage vector updated at the present time step and $\mathbf{V^P_{ki}} = [\mathbf{V^P_k}, \mathbf{V^P_i}]$ is the voltage vector obtained from the previous time step. In the outer loop, the deviation of the interchange power between subsystems k i can then be evaluated by $\Delta \mathbf{s_{ki}} = \mathbf{Y_{ki}} \cdot \mathbf{V_{ki}} \circ \overline{\mathbf{V}}_{\mathbf{ki}} - \mathbf{Y_{ki}} \cdot \mathbf{V^P_{ki}} \circ \overline{\mathbf{V}}^P_{\mathbf{ki}}$. A threshold ϵ_o is set for the power error in the outer loop iteration. Assume L_o is the counter of the iteration numbers of the outer loop, and $Iter_o^{max}$ is a given upper limit of the outer loop's number of iterations. Satisfying one of the following stopping criteria will be sufficient to end the overall iterative process:

$$\Delta \mathbf{s_{ki}} \leq \epsilon_o, \tag{5.47}$$

$$L_o > Iter_o^{max}, \tag{5.48}$$

Distributed Algorithm

The neighboring subsystems are electrically coupled by the powers flowing through the tie lines, which are $\mathbf{S^I_{ext}}$ (see Figure 5.10), $\mathbf{S^I_{kk}}$, and $\mathbf{S^I_{jj}}$ (see (5.42)). These power exchanges can be determined by the voltages across the interfaces between the subsystem and its neighbors. For a better explanation, we use the jth passive subsystem as an example. Assume $\mathbf{Y_{ji}}$, which can be derived via (5.30), is the admittance matrix of the tie lines linking subsystems j and i; and assume $\mathbf{V_{ji}} = [\mathbf{V_j}, \mathbf{V^P_i}]$ is the voltage vector. As subsystem j is represented by the power flow equations only, the power exchange can be updated as follows:

$$\begin{cases} \mathbf{Y_{jj}} \cdot \mathbf{V_{jj}} \circ \overline{\mathbf{V}}_{\mathbf{jj}} - \mathbf{S^L_{jj}} - \mathbf{S^I_{jj}} = 0 \\ \mathbf{S^I_{jj}} = \mathbf{Y_{ji}} \cdot \mathbf{V_{ji}} \circ \overline{\mathbf{V}}_{\mathbf{ji}}. \end{cases} \tag{5.49}$$

The inner-loop iteration for the jth subsystem is explained as follows. First, the interface of the neighboring ith subsystem is treated as a reference bus, meaning its voltage $\mathbf{V^P_i}$ obtained from the past history will keep constant during the computations of power flow (or state reachable sets) in the subsystem j. $\mathbf{V_j}$ is then obtained after the computation of power flow (or state sets) in the jth subsystem finishes.

Let $\Delta \mathbf{V_{jj}}$ represent the differences of subsystem j voltages obtained from two iterations. ϵ_i is the threshold set for the inner-loop iteration. L_i is the number of iterations of the inner loop. $Iter_i^{max}$ is the given upper limit of the inner loop's number of iterations. The stopping criterion for the inner-loop iterations is that any of the following conditions are met:

$$\Delta \mathbf{V_{jj}} \leq \epsilon_i, \tag{5.50}$$

$$L_i > Iter_i^{max}. \tag{5.51}$$

The power exchanges between subsystems j and i are calculated at each iteration as soon as $\mathbf{V_j}$ in subsystem j are updated. Therefore, electrical laws (e.g., Ohm's law) which determines the performance of the interfaces between subsystems are strictly followed in each time interval of the inner-loop iterations.

Data Communication between Subsystems

Considering that both power flow and reachable set calculations in subsystems take interface quantities at each time step, it is important to design an effective data communication scheme between subsystems to enable the implementation of DFA. For a subsystem, the number of iterations and computation time needed for the inner-loop calculations are unknown to its neighboring subsystems. A status "Flag" is then introduced to exchange the computation status and results between subsystems. The first item in the "Flag" is *Subsystem-ID*, that is, the ID of the neighboring subsystems. Another item is a binary indicator *Convergence* to show whether the subsystem is converged (if *Convergence* =1) or not yet. Once the iteration ends, the voltages at the interface will be stored in item *Results*. In summary, the "Flag" is given in (5.52):

$$Flag = [Subsystem\text{-}ID, Convergence, Results].   (5.52)$$

The status "Flag" technique is adopted from [24]. In practice, if the NM coordinator is built upon a software-defined networking architecture, the "Flag" can be easily implemented through the OpenFlow controller, as mentioned in Chapter 4. The status flag technique helps achieve the following objectives:

- Each subsystem is guaranteed to receive the newly converged data from its neighboring subsystems.
- Only interface data are transferred, which means the design preserves privacy of subsystems. Because a very small amount of data are to be exchanged, it is feasible to achieve real-time data encryption to help protect against common cyber-attacks such as man-in-the-middle attacks.
- Programming the "Flag" is flexible, allowing new attributes to be added and obsolete features removed as needed.

The essential idea of data exchange is illustrated in Figure 5.12.

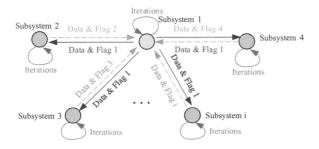

Figure 5.12 Data exchange between subsystems.

There are three steps involved within one time step:

- The information of each subsystem's interface, that is, S_{kk}^I or S_{jj}^I in (5.42), is updated and shared with other subsystems.

  Specifically, in distributed power flow calculations, the interface information is crisp-value-based data, whereas calculating the subsystems' reachable sets involves set-based data. Status flags are sent to neighboring subsystems, which can be used to verify the shared information.

- As soon as all the interface data needed by one subsystem are available, the inner-loop (power flow or reachable state set) iterations will begin.

  For instance, once the interface data and status flags from the neighbors of Subsystem 1 (Subsystems 2–4, and i; see Figure 5.12) are received, the inner-loop calculation for Subsystem 1 will start. The interface data obtained from latest iterations of the neighboring subsystems will be the inputs for the power flow or reachable sets calculations.

- After all subsystems finish the inner-loop calculations, the status flags including interface data are broadcast to initiate the next outer-loop iteration.

5.6.3   Implementation of DQG

In this chapter, DQG is combined with reachable set calculation to efficiently assess the NM systems' stability margin. Specifically, eigenvalues at the vertices of the reachable set of a subsystem are estimated via (5.23) and (5.24), which will give the Geršgorin disks for the ith subsystem (i.e., s_{kk} and $r_k(S_{i,P})$, see (5.23)). The stability of the ith subsystem can be quickly verified by checking the largest real part of the estimated eigenvalues $\alpha_{i,max}^e$, which are calculated as follows:

$$\alpha_{i,max}^e = max(s_{kk} + r_k(S_{i,P})). \tag{5.53}$$

5.6.4   Stability Margin Assessment

The stability margin of a networked microgrids system subject to disturbances can be established using the procedures detailed in Subsection 5.3.2. The keystone techniques are the reachable set calculation and Geršgorin disks computation discussed earlier.

5.7   Testing and Validation of FA and DFA

The effectiveness of the presented FA and DFA in combination with the quasi diagonalized Geršgorin approach is verified on the test NM system with six microgrids, as illustrated in Figure 5.13. We run the NM system in the islanded mode in order to better capture the impact of uncertainties. Microgrid 1 is powered by a small conventional generator, which is modeled by a classical synchronous generator to control the frequency and voltage of the NM system. The rest of microgrids are power electronic

Table 5.1 Parameters for inverter controllers in microgrids shown in Figure 5.13

Microgrids	Parameters					
	T_f	T_r	K_P	T_P	K_Q	T_Q
2	0.01	0.01	0.45	0.02	0.45	0.02
3	0.01	0.01	0.80	0.01	0.80	0.01
4	0.01	0.01	0.50	0.02	0.50	0.02
5	0.01	0.01	0.30	0.02	0.30	0.02
6	0.01	0.01	0.40	0.02	0.40	0.02

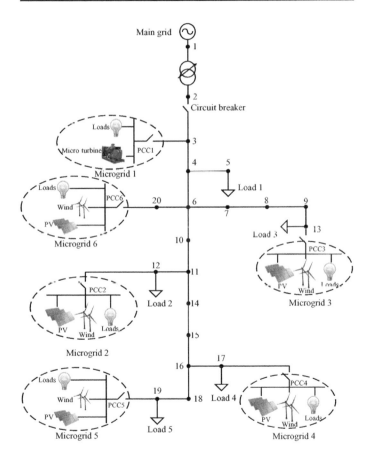

Figure 5.13 A typical NM system.

dominant, which are powered by DERs with inverters as interfaces. Table 5.1 summarizes the parameters for microgrid controllers. The parameters of the backbone system can be found in [25].

For the NM system operated in the islanded mode, the dimensions of its extended admittance matrix $\overline{\mathbf{Y}}$ in Figure 5.13 are 36×18. Correspondingly, the dimensions of the bus voltage vector $\mathbf{V}$, the extended bus voltage vector $\overline{\mathbf{V}}$, and power vector $\overline{\mathbf{S}^G}, \overline{\mathbf{S}^L}$ are 18×1, 36×1, 36×1, and 36×1, respectively.

The FA and DFA algorithms are implemented in the COntinuous Reachability Analyzer (CORA) [26] environment. A time step of 0.010 s is selected for the reachability analysis.

5.7.1   Reachable Set Calculation in FA

Reachable Set Calculation

We consider a case with different levels of active power fluctuation in Microgrid 6, assuming the power output fluctuates around the baseline power by $\pm 5\%$, $\pm 10\%$, $\pm 15\%$, and $\pm 20\%$.

We examine the reachable sets for two key state variables X_{pi} (the state variable in the upper proportional-integral (PI) control block shown in Figure 5.14) and X_{qi} (the state variable in the lower PI controller shown in Figure 5.14). Figures 5.15–5.17 respectively illustrate the state sets of X_{pi} and X_{qi} along the time in three microgrids, 6, 2, and 5, under the four levels of uncertainties.

The following can be observed:

- Through reachable set calculation, we can directly obtain the possible operation range of an NM system under disturbances.
- As the uncertainty level increases, the areas of zonotopes along reachable sets increase accordingly. The overapproximation of reachable sets has been validated by comparing with time-domain simulations.
- The reachtubes associated with Microgrid 6 converge along the timeline, rather than consistently increase, as shown in Figure 5.15. A major reason is that Microgrid 6 is electrically close to Microgrid 1, which is powered by the synchronous generator.

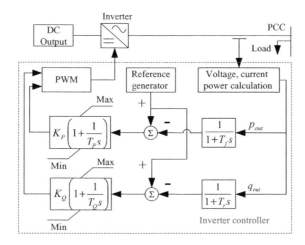

Figure 5.14 Controller of DERs in NMs.

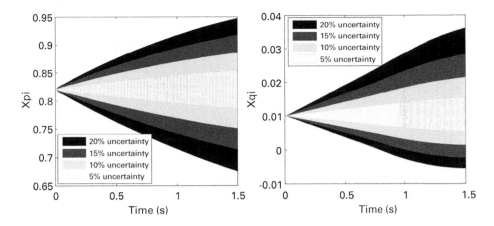

Figure 5.15 Reachable set in Microgrid 6 projected to the timeline.

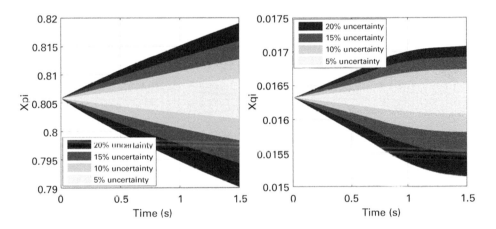

Figure 5.16 Two-dimensional projection of the reachable set in Microgrid 2.

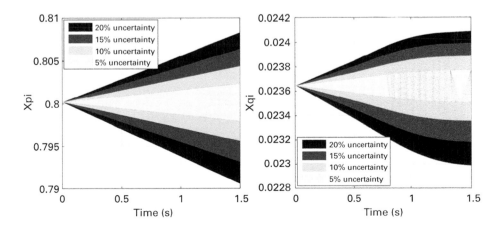

Figure 5.17 Two-dimensional projection of the reachable set in Microgrid 5 projected.

Therefore, the inertia of Microgrid 1 helps reduce the effect of uncertainties on Microgrid 6.

- The impact of the uncertainties in Microgrid 6 on Microgrid 2 and Microgrid 5 are different. We can see that Microgrid 5 is less affected than Microgrid 2, mainly because Microgrid 5 has a shorter electrical distance to Microgrid 6 than Microgrid 2 does.

Reachable Set Verification in FA

The efficacy of FA is validated by comparing with the repeated time-domain simulations. Here, in order to clearly illustrate the comparison, only ten time-domain trajectories are selected to compare against the reachable sets generated by FA. Figures 5.18 and 5.19 shows the comparison results for Microgrid 2 and Microgrid 5, respectively. From these two figures, we can observe the following:

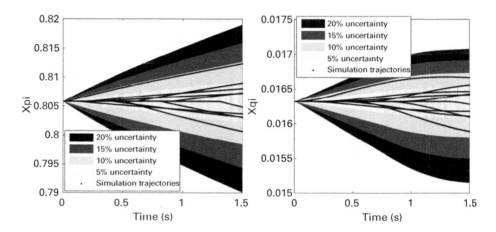

Figure 5.18   Validation of the reachable set in Microgrid 2 through time-domain simulations.

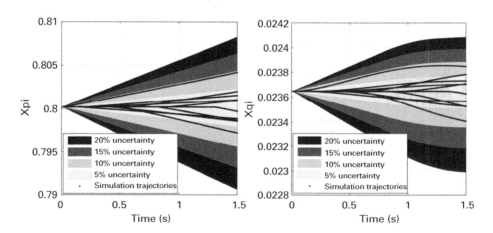

Figure 5.19   Validation of the reachable set in Microgrid 5 through time-domain simulations.

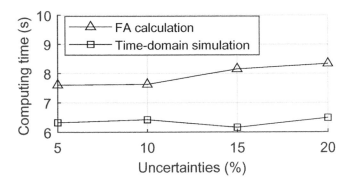

Figure 5.20 Computing time for analyzing 1.5 s dynamics.

- The time-domain trajectories are fully enclosed by the reachable sets, meaning FA computes a sound overapproximation of the possible dynamic behaviors of the system. The overapproximation property is highly desirable, as it allows a formal verification of the NM system functionalities (e.g., fault ride-through) for guaranteed and provably reliable NM operations.
- However, for a large-scale nonlinear NM system, certain factors may lead to an overly conservative approximation of the reachable states. New optimization-based approaches such as the continuous Lagrangian reachability (CLRT) [27] might be a potential solution to make the computed reachable sets as tight as possible.

Efficiency of FA

The computation time for the ten runs of time-domain simulations versus that for FA have been compared and summarized in Figure 5.20. Both calculations are performed on a 3.4 GHz PC. The results validate that the efficiency of FA meets the needs for the formal verification of NM dynamics under uncertainties.

Simulation Step Size Selection

Simulation step size will affect FA accuracy because the NM is a typical nonlinear system. Given a ±20% active power uncertainty in Microgrid 6, the reachable sets X_{pi} of Microgrid 6 obtained with five step sizes are compared in Figure 5.21. As can be seen from Figure 5.21:

- More accurate results can be obtained when a smaller time step 0.001 s is adopted. However, as shown in Figure 5.22, the drawback is that the computational cost of FA is much higher in this case.
- The computational time can be decreased by increasing the time step, whereas the accuracy of FA inevitably drops, as shown in Figure 5.22.
- When we set the time step as 0.015 s, after 15 iterations, the simulation process suspends. The major reason is that the matrix $\mathbf{G_y}$ is close to singular; thus, the corresponding results may be inaccurate.

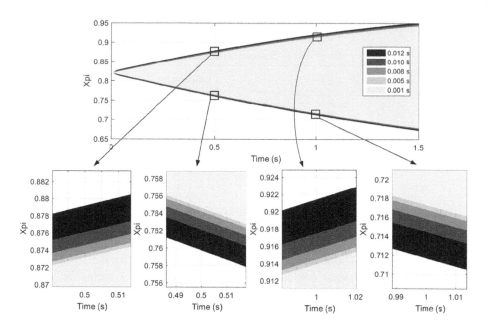

Figure 5.21 The effect of time step sizes on the FA results.

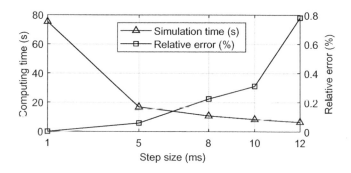

Figure 5.22 Calculation time and relative errors using different simulation step sizes.

- When we keep increasing the time step, the simulation process suspends much more quickly. This is particularly evident as a step size of 0.05 s is adopted – only after three simulation steps is the FA process aborted.
- When we assume the result of a 0.001 s time step is accurate, Figure 5.22 summarizes the relative errors of the other time steps at 1.0 s.
- A compromise between accuracy and efficiency needs to be made. As shown in Figure 5.22, a step size of 0.010 s can be a more reasonable choice than either an overly small step size of 0.001 s (more accurate but too expensive) or a larger step of 0.012 s (accelerated FA with deteriorated accuracy).

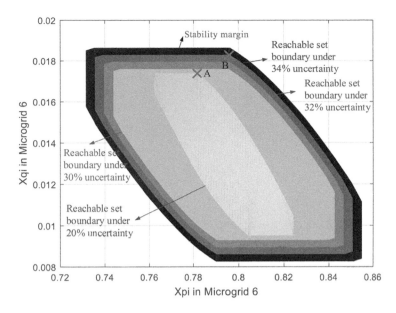

Figure 5.23 Process of assessing the stability margin of Microgrid 6 at $t = 0.5$ s.

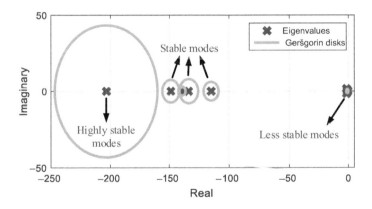

Figure 5.24 Geršgorin disks and eigenvalues associated with vertex A.

5.7.2   Assessment of Stability Margin through FA Enhanced the Quasi diagonalized Geršgorin Technique

Assessment of Stability Margin

The purpose of this case is to show how FA and the quasi diagonalized Geršgorin theorem are combined to evaluate the stability margins along the timeline. Figure 5.23 gives the stability margin of Microgrid 6 obtained at 0.5 s. The estimated Geršgorin disks corresponding to vertices A and B in Figure 5.23 are illustrated in Figures 5.24 and 5.27. Exact eigenvalues are provided as well for a reference.

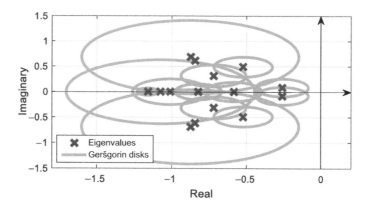

Figure 5.25 Zoomed-in Geršgorin disks around the imaginary axis in Figure 5.24.

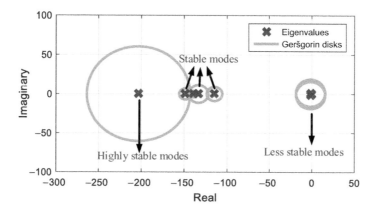

Figure 5.26 Geršgorin disks and eigenvalues associated with vertex B.

Based on Figures 5.23–5.26, the following can be seen:

- We can efficiently obtain the stability margin, which verifies that the integration of FA and a quasi diagonalized Geršgorin technique is feasible.
- The quasi diagonalized Geršgorin approach is particularly useful in saving the computational efforts of the exact eigenvalue calculations when the NM system is operating fairly distant from its stability margin, such as operating at point A in Figure 5.23, while still giving correct stability information.
- If the NM system is close to its stability boundary, the results of a quasi diagonalized Geršgorin will be conservative. For example, Figure 5.26 shows the stability results at point B. In this case, we need to calculate the exact eigenvalue to further inspect the system's stability.
- Figures 5.24 and 5.26 demonstrate that the NM system has three types of modes, including the so-called "highly stable modes," "stable modes," and "less stable modes." Since the system's dynamics are dominated by the eigenvalues of the less stable modes, computational efforts should be focused on checking the Geršgorin

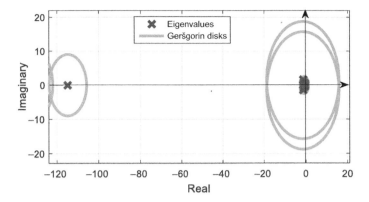

Figure 5.27 Zoomed-in Geršgorin disks around the imaginary axis in Figure 5.26.

disks in the "less stable" zones. Figures 5.25 and 5.27 provide magnified illustrations of those disks and corresponding eigenvalues.

Efficiency of FA Enhanced with the Quasi diagonalized Geršgorin Technique

As a quick recap of Figure 5.8, the stability margin can be probed by repeatedly performing eigenvalue estimations, mainly through the quasi diagonalized Geršgorin technique, until the stability criterion (i) is no longer satisfied. For this test NM system, if the exact eigenanalysis is performed at each time step, it would take 29.8990 s to identify the stability margin. In contrast, it only takes 17.1653 s to establish the stability margin when the quasi diagonalized Geršgorin technique is adopted during the FA process. This means the calculation time needed for the Geršgorin-enhanced FA is only 57.41% of that used for the FA combined with the exact eigenanalysis. Therefore, the quasi diagonalized Geršgorin-based FA is an efficient method for the formal verification of the NM system stability subject to uncertainties.

Use of FA for NM Operations

How to reliably assess an NM system's stability is one of the operator's primary concerns with operating the system. Since FA can improve the situational awareness and controllability of an NM system, FA enables NMs to become dependable resources that can enhance the distribution grid resiliency. System operators can leverage the stability margins obtained by FA in several ways:

- Given accurate nowcast and forecast of loads and DER outputs, operators can now monitor and understand the possible dynamical behaviors of an NM system even under high levels of DER penetration.
- Preventive or corrective control or dispatch strategies can be performed correctly when the NM operating states are in close proximity to the system's stability margin. Thus, the resiliency of the NM system can be improved owing to the FA's capability of measuring the proximity of the operating states to the stability margin.

- With the FA tool, system operators can localize those critical control parameters or components by examining trajectory sensitivities of the NM system. This information can provide important measures for optimized tuning or reconfiguration to maximize the stability and resiliency of an NM system.

5.7.3   DFA with System Decomposition

Two different partitions are performed in the FA analysis of the test NM system to show the effectiveness of the decomposition technique.

Test A: Bipartite NM System

In this case, power flow analysis is performed on a bipartite NM system whose partition has only active subsystems without any passive ones, that is, $N = 2$ and $M = 0$ in (5.42). Lines 6 through 10 are the cut that partitions the NM system into two disjoint subsystems, Subsystem 1 consisting of Microgrids 1, 3, and 6, and Subsystem 2 including Microgrids 2, 4, and 5. The power flow calculation of Subsystem 1 takes the power flowing from node 10 as an input. Likewise, the power flowing from node 6 is an input for the power flow analysis of Subsystem 2.

A tolerance ϵ_i of $1.0e - 10$ is selected for the power flow analysis of Subsystems 1 and 2. Meanwhile, a tolerance ϵ_o of $1.0e - 5$ is used for the power flow iterations at interface. As the value changes in p.u. are very small during iterations, an $L2$ norm is taken for the actual error v_i [28], which is further scaled up to magnify changes in the power flow iterations, resulting in an index $\|r_i\|_2$ as follows:

$$\|r_i\|_2 = -10/ln(\|v_i\|_2). \tag{5.54}$$

The voltage magnitude comparison is given in Figure 5.28 to show the difference between the distributed and centralized solutions. Based on these results, the two

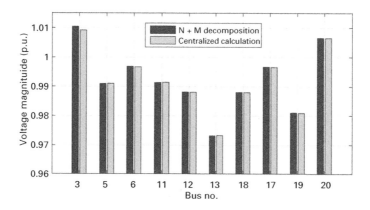

Figure 5.28 Test A: voltages obtained from the $N+M$ decomposition and the centralized calculation.

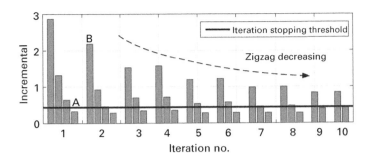

Figure 5.29 Test A: Convergence process of power flow solution for Subsystem 1.

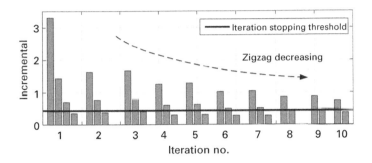

Figure 5.30 Test A: Convergence process of power flow solution for Subsystem 2.

approaches give nearly identical results, showing the efficacy of distributed power flow based on the $N + M$ decomposition.

The changes in variables at the Newton iterations in Subsystems 1 and 2 are shown in Figures 5.29 and 5.30, respectively. From the results, we can see the convergence for Subsystems 1 or 2 has a zigzag process. This nonmonotonic convergence is due to the fact that power flow calculations for each subsystem rely on the interface flow (S_{kk}^I and S_{jj}^I in (5.42)) obtained from the previous iteration. As a result, a subsystem, which reaches convergence at the current step (e.g., point A in Figure 5.29), may have a large mismatch again after the interface data are exchanged between subsystems. The Newton iteration, therefore, has to start over again to bring down the newly introduced mismatch, such as point B shown in Figure 5.29. Different subsystems normally have different subiteration processes. For example, during the second iteration, it takes four subiterations for Subsystem 1 to converge whereas it takes three subiterations for Subsystem 2 to finish its inner loop calculation. Thus, it is necessary to use status flags to keep synchronous update between subsystems.

The mismatches in the power flows through the interface between two consecutive iterations are demonstrated in Figure 5.31. Based on these results, we can see that the convergence process at the subsystem interface is monotonic, indicating a desirable global convergence. It further verifies the efficacy of $N+M$ decomposition in enabling the distributed NM calculation.

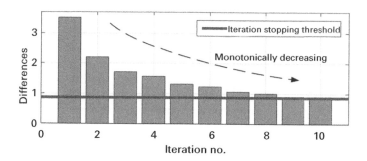

Figure 5.31 Test A: Convergence process at the subsystem interface.

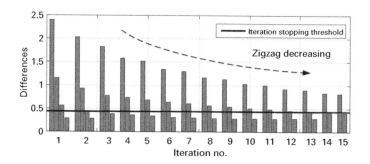

Figure 5.32 Test B: Convergence process of power flow solution for Subsystem 1.

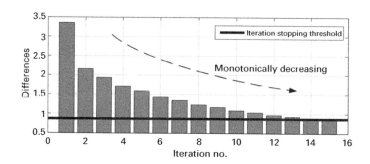

Figure 5.33 Test B: Convergence process at the subsystem interface.

Test B: Quadripartite NM System

In this case, the NM system is further decoupled into a quadripartite network in order to gain more insight in a more distributed NM system as compared with the bipartite network in Test A. Lines 6–7, 6–10, and 15 and 16 form a cut set that partitions the NM system into four disjoint subsystems. The system settings are otherwise identical to those in Test A.

Figure 5.32 illustrates the changes of variables at iterations in Subsystem 1, which now consists of Microgrids 1 and 6. The mismatches in the power flows through the subsystem interface between two consecutive iterations are shown in Figure 5.33.

From Figures 5.32 and 5.33 and the comparison between Tests A with II, we can gain the following insights:

- If the NM system is partitioned into more subsystems, less calculation time is needed to complete an inner-loop iteration for a subsystem. In Test A, it takes 0.05 s to finish the four subiterations at the first iteration step in Figure 5.29. In contrast, it takes 0.03 s in Test B to finish a similar process in Figure 5.32, meaning Test B is 40% more efficient in performing one iteration.
- Making overly granular partitions, which means a k-partite NM system with a large k, the number of outer-loop iterations might increase significantly. As can be observed from Figures 5.29 and 5.33, Test A requires ten iterations to reach the global converge while Test B needs 15 iterations for the same purpose. One major reason is that we have more frequent data exchange between subsystems in Test B.

5.7.4   DFA for Calculating Reachable Set

The same bipartite NM system in Test A is adopted in this test. Five levels of active power fluctuations, that is, $\pm 1\%$, $\pm 5\%$, $\pm 8\%$, $\pm 10\%$, and $\pm 12\%$ around the baseline power output, are set for Microgrid 2.

Reachable Set for NMs
Figures 5.34 and 5.36 respectively illustrate the 2D time-domain projections for the reachable set of Microgrids 3 and 4, where the y-axes represent X_{pi} (the state variable of active power control) and X_{qi} (the state variable of reactive power control). Figure 5.35 shows cross sections zoomed in at 0.2 s and 0.5 s. More details of X_{pi} and X_{qi} can be found in Subsection 5.7.1.

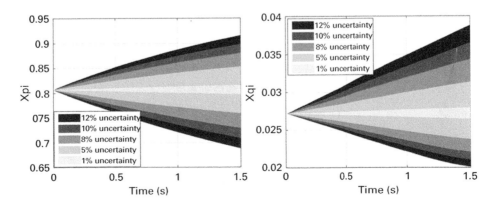

Figure 5.34 Two-dimensional projections of reachable sets X_{pi}/X_{qi} in Microgrid 3 of Subsystem 1.

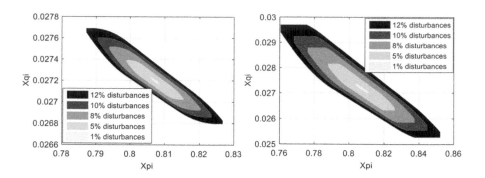

Figure 5.35 Zoomed-in reachable sets at 0.2 s and 0.5 s in Figure 5.34.

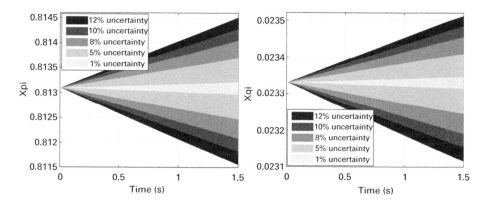

Figure 5.36 Two-dimensional projections of reachable sets X_{pi}/X_{qi} in Microgrid 4 of Subsystem 2.

From Figures 5.34 and 5.36, we can see the following:

- Operational boundaries of the test NM system corresponding to different uncertainty levels can be obtained via DFA. The $N + M$ decomposition approach is therefore feasible and enables effective DFA calculation.
- The magnified plot in Figure 5.35 indicates that, when the uncertainty level increases, the size of reachable sets will increase. We can further verify the accuracy of the DFA calculation by comparing the reachable sets obtained from the DFA with those from the centralized FA.
- Figures 5.34 and 5.36 show that the reactive power output of a microgrid may be affected by the fluctuations in active power mainly owning to the feeder resistances in the backbone network. As an example, the deviations of X_{pi} and X_{qi} at 1.5 s are summarized in Figure 5.37.
- Because DFA gives the reachable sets caused by different levels of disturbances, it can be used to estimate importance measures that can identify those critical disturbances that threaten the NM stability performance most. Such information can also be useful for the efficient estimation of the stability margin of NMs subject to uncertainties.

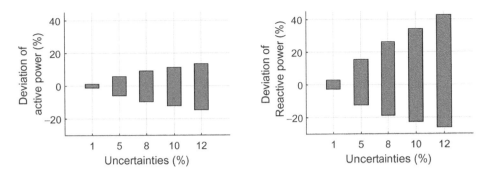

Figure 5.37 Deviations in active and reactive power.

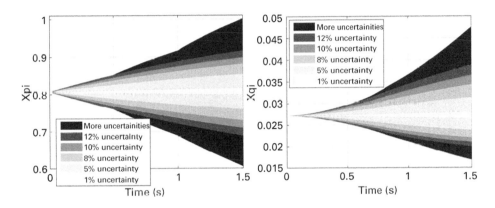

Figure 5.38 Microgrid 3 reachable sets X_{pi} and X_{qi} under more uncertainties.

Effect of Extreme DERs Injections on NMs

To observe the impact of extreme DER fluctuations on the NM system dynamics, a $\pm 20\%$ disturbance is applied to perturb the baseline active power output of Microgrid 2 at 0.5 s, followed by a $\pm 30\%$ disturbance started at 1.0 s. Figure 5.38 shows the 2D projections of reachable sets for Microgrid 3. As can be seen, when the NM system is subject to large DER disturbances, the reachtube's cross-section area significantly increases, showing large deviations from the initial point and a tendency of divergence. A high-level integration of DERs without proper coordination thus can worsen the stability of the NM system. By evaluating severe cases, we can pinpoint critical disturbances through reachable set results. Once such potential hazards are detected, predictive or corrective control actions can be triggered immediately to ensure the NM system is brought back to its stability region.

Time-Domain Simulation for Validating DFA

The capability of DFA in enclosing possible scenarios is validated through time-domain simulations. Under the same uncertainty level, a finite number of time-domain trajectories are generated and compared against the DFA results. Figures 5.39 and 5.40 demonstrate the simulation curves of selected X_{pi} and X_{qi} in Subsystem 2.

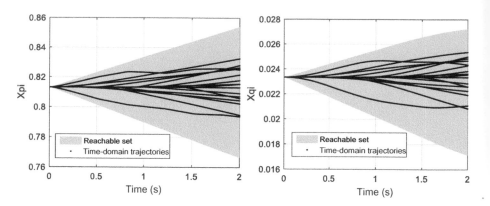

Figure 5.39 Time-domain simulation verification in Microgrid 4.

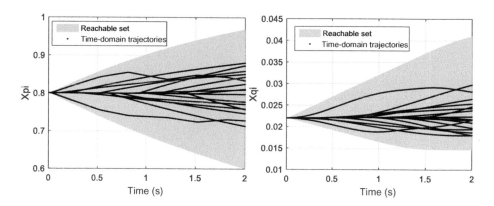

Figure 5.40 Time-domain simulation verification in Microgrid 5.

From Figures 5.39 and 5.40, the following can be observed:

- All time-domain trajectories of the selected states are completely enclosed by the corresponding reachtubes of the NM system. It is therefore demonstrated that DFA provides the overapproximation of the reachable states of the NM system. DFA is also found to produce identical reachable set results to those generated by the centralized FA.
- This case gives a reasonable level of overapproximation useful for checking the safety property of an NM system. However, it is still an open problem that a direct DFA calculation might lead to overly conservative results due to the strong nonlinearity in microgrids, switching operations and interfacing errors. Even though there exist a few techniques for set splitting [8] and bounds tightening [29], finding tight reachtubes for large nonlinear or hybrid dynamical systems such as the NM system continues to be a subject undergoing intense study.

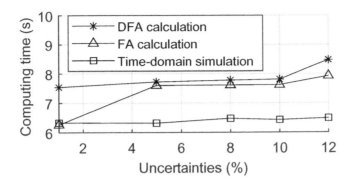

Figure 5.41 Computing time needed for 1.5 s of dynamic analysis.

Efficiency in DFA

The computational time needed for one run of DFA and FA is compared with that for ten runs of time-domain simulation, as shown in Figure 5.41. One can see the following:

- DFA demonstrated competitive performance in terms of reachable set calculation and stability analysis.
- Since DFA is formal, one DFA calculation is able to overapproximate all possible time-domain trajectories caused by the predefined set-based uncertainties. On the other hand, it will be prohibitively expensive for the time-domain simulation to enumerate the infinitely many scenarios. For formal verification purposes, DFA is theoretically a better option over the traditional time-domain simulations.
- The information exchange between subsystems adds extra cost in DFA, making it slightly slower than FA in this small-scale case. For a complex NM system with a large number of nodes (n), DFA would outperform FA due to an $O(n^5)$ complexity of reachable set calculation [6], where the gain of computation efficiency in subsystems would greatly outweigh the communication cost involved.

Evolution of Reachable Sets at Iterations

Figures 5.42 and 5.43 demonstrate how reachable sets of Microgrid 3 evolve within the iteration process at each discrete step. Figure 5.42 shows 2D snapshots of X_{pi} and X_{qi} at 0.2 s, and Figure 5.43 illustrates the 2D reachable sets at 1.0 s.

- After several iterations, we can obtain converged reachable sets in subsystems.
- Through parallel iterations, we can calculate reachable sets. This feature could enable efficient distributed stability assessment of large NM systems as the calculation process is particularly suited for the plug-and-play of microgrids and/or microgrid components.

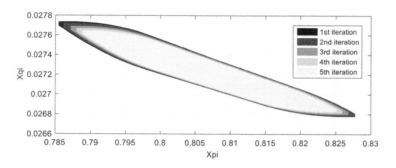

Figure 5.42 Reachable set iterations at 0.2 s.

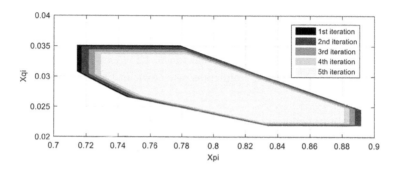

Figure 5.43 Reachable set iterations at 1.0 s.

5.7.5   DQG-Based DFA Approach to Probing the Stability Margin

The feasibility of DQG-empowered DFA in probing the stability margin at a specific time point is also demonstrated in this case. Microgrid 5's stability margin at 0.2 s is illustrated in Figure 5.44. The stability performance of three operating points A, B, and C marked in Figure 5.44 are elaborated in Figures 5.45–5.47, respectively, through Geršgorin disks. The test results show the following:

- We can obtain the stability margin via DQG-enabled DFA. The DFA empowered by DQG seems to be a feasible and practical approach for stability assessment purposes.
- When the NM system is operating in a fairly stable point, such as point A in Figure 5.44, DQG serves as an efficient approach to confirm the system stability, avoiding the expensive full-scale evaluation of exact eigenvalues.
- The centers of Geršgorin disks will become larger and move toward the instability region when the uncertainty level is increased. This can be observed by checking the Geršgorin results in Figure 5.46 corresponding to point B in Figure 5.44.
- Part of the Geršgorin disks from DQG calculation would enter the instability zone when the NM system approaches the boundary of stability. Under this condition, the DQG results are no longer meaningful for the determination of the NM system

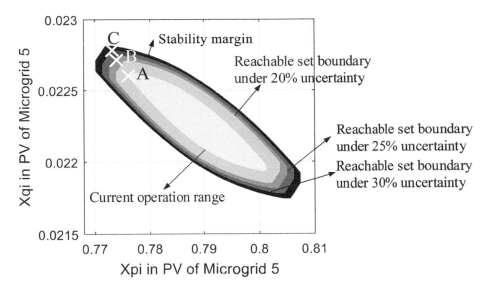

Figure 5.44 Shape of Microgrid 5's stability margin at 0.5 s.

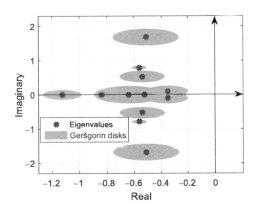

Figure 5.45 Locations of Geršgorin disks/eigenvalues evaluated at A.

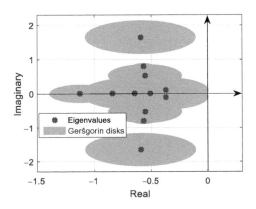

Figure 5.46 Locations of Geršgorin disks/eigenvalues evaluated at B.

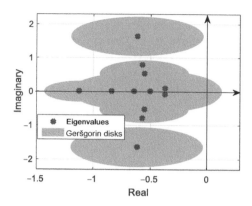

Figure 5.47 Locations of Geršgorin disks/eigenvalues evaluated at C.

stability, as shown in Figure 5.47 corresponding to point C in Figure 5.44. Thus, a full-scale estimation of exact eigenvalues becomes the only option.

References

[1] Y. Li, P. Zhang, and P. B. Luh, "Formal Analysis of Networked Microgrids Dynamics," *IEEE Transactions on Power Systems*, vol. 33, no. 3, pp. 3418–3427, 2018.

[2] A. El-Guindy, K. Schaab, B. Schürmann, D. Han, O. Stursberg, and M. Althoff, "Formal LPV Control for Transient Stability of Power Systems," in *Proc. of the IEEE PES General Meeting*, pp. 1–5, 2017.

[3] A. El-Guindy, Y. C. Chen, and M. Althoff, "Compositional Transient Stability Analysis of Power Systems via the Computation of Reachable Sets," *American Control Conference (ACC), 2017*. IEEE, pp. 2536–2543, 2017.

[4] X. Jiang, Y. C. Chen, and A. D. Domínguez-García, "A Set-Theoretic Framework to Assess the Impact of Variable Generation on the Power Flow," *IEEE Transactions on Power Systems*, vol. 28, no. 2, pp. 855–867, 2013.

[5] S. Chong, J. Guttman, A. Datta, et al., "Report on the NSF Workshop on Formal Methods for Security," *arXiv preprint arXiv:1608.00678*, 2016.

[6] M. Althoff, "Formal and Compositional Analysis of Power Systems Using Reachable Sets," *IEEE Transactions on Power Systems*, vol. 29, no. 5, pp. 2270–2280, 2014.

[7] M. Altho, "Reachability Analysis and Its Application to the Safety Assessment of Autonomous Cars," Ph.D. dissertation, Technische Universität München, 2010.

[8] M. Althoff, O. Stursberg, and M. Buss, "Reachability Analysis of Nonlinear Systems with Uncertain Parameters Using Conservative Linearization," *47th IEEE Conference on Decision and Control, 2008. CDC 2008*. IEEE, pp. 4042–4048, 2008.

[9] T. Ding, R. Bo, F. Li, et al., "Interval Power Flow Analysis Using Linear Relaxation and Optimality-Based Bounds Tightening (OBBT) Methods," *IEEE Transactions on Power Systems*, vol. 30, no. 1, pp. 177–188, 2015.

[10] M. Althoff and B.-H. Krogh, "Reachability Analysis of Nonlinear Differential-Algebraic Systems," *IEEE Transactions on Automatic Control*, vol. 59, no. 2, pp. 371–383, 2014.

[11] I. J. Pérez-Arriaga, G. C. Verghese, and F. C. Schweppe, "Selective Modal Analysis with Applications to Electric Power Systems, Part I: Heuristic Introduction," *IEEE Transactions on Power Apparatus and Systems*, no. 9, pp. 3117–3125, 1982.

[12] C. Michailidou and P. Psarrakos, "Geršgorin Type Sets for Eigenvalues of Matrix Polynomials," *Electronic Journal of Linear Algebra*, vol. 34, no. 1, pp. 652–674, 2018.

[13] F. Feng and P. Zhang, "Implicit Z_{bus} Gauss Algorithm Revisited," *IEEE PES Letters, IEEE Transactions on Power Systems*, vol. 35, no. 5, pp. 4108–4111, 2020.

[14] F. Feng and P. Zhang, "Enhanced Microgrid Power Flow Incorporating Hierarchical Control," *IEEE PES Letters, IEEE Transactions on Power Systems*, vol. 35, no. 3, pp. 2463–2466, 2020.

[15] J. H. Wilkinson, *The Algebraic Eigenvalue Problem*. Clarendon Press Oxford, 1965.

[16] H.-D. Chang, C.-C. Chu, and G. Cauley, "Direct Stability Analysis of Electric Power Systems Using Energy Functions: Theory, Applications, and Perspective," *Proceedings of the IEEE*, vol. 83, no. 11, pp. 1497–1529, 1995.

[17] N. Duan and K. Sun, "Power System Simulation Using the Multistage Adomian Decomposition Method," *IEEE Transactions on Power Systems*, vol. 32, no. 1, pp. 430–441, 2017.

[18] J. H. Chow, "Time-Scale Separation in Power System Swing Dynamics: Singular Perturbations and Coherency," *Encyclopedia of Systems and Control*, SpringerReference, pp. 1465–1469, 2015.

[19] R. J. Sánchez-García, M. Fennelly, S. Norris, et al. "Hierarchical Spectral Clustering of Power Grids," *IEEE Transactions on Power Systems*, vol. 29, no. 5, pp. 2229–2237, 2014.

[20] P. Zhang, J. Marti, and H. Dommel, "Network Partitioning for Real-Time Power System Simulation," *International Conference on Power System Transients, Montreal, Canada*, pp. 1–6, 2005.

[21] M. Crow and M. Ilic, "The Parallel Implementation of the Waveform Relaxation Method for Transient Stability Simulations," *IEEE Transactions on Power Systems*, vol. 5, no. 3, pp. 922–932, 1990.

[22] C. Wang, Y. Li, K. Peng, B. Hong, Z. Wu, and C. Sun, "Coordinated Optimal Design of Inverter Controllers in a Micro-Grid with Multiple Distributed Generation Units," *IEEE Transactions on Power Systems*, vol. 28, no. 3, pp. 2679–2687, 2013.

[23] Y. Zhou, P. Zhang, and M. Yue, "An ODE-Enabled Distributed Transient Stability Analysis for Networked Microgrids," *2020 IEEE Power and Energy Society General Meeting (PESGM)*, pp. 1–5, 2020.

[24] Z. Zhang, Y. Cheng, S. Nepal, D. Liu, Q. Shen, and F. Rabhi, "A Reliable and Practical Approach to Kernel Attack Surface Reduction of Commodity OS," *arXiv preprint arXiv:1802.07062*, 2018.

[25] S. Papathanassiou, N. Hatziargyriou, K. Strunz, "A Benchmark Low Voltage Microgrid Network," *Proceedings of the CIGRE Symposium: Power Systems with Dispersed Generation*, pp. 1–8, 2005.

[26] M. Althoff, "CORA 2016 Manual," http://archive.www6.in.tum.de/www6/pub/Main/SoftwareCORA/Cora2016Manual.pdf, 2016.

[27] J. Cyranka, M. A. Islam, S. A. Smolka, S. Gao, and R. Grosu, "Tight Continuous-Time Reachtubes for Lagrangian Reachability," *2018 IEEE Conference on Decision and Control (CDC)*. IEEE, pp. 6854–6861, 2018.

[28] C. Robert, "Machine Learning, a Probabilistic Perspective," 2014.

[29] I. Araya and V. Reyes, "Interval Branch-and-Bound Algorithms for Optimization and Constraint Satisfaction: A Survey and Prospects," *Journal of Global Optimization*, vol. 65, no. 4, pp. 837–866, 2016.

6   Active Fault Management for Networked Microgrids

6.1   Introduction

A key to promoting interconnection among networked microgrids is that networked microgrids should systematically support grid resiliency under grid disturbances rather than negatively impacting the disturbed grid. In this chapter, we develop a new concept of microgrid active fault management (AFM). By integrating distributed optimization and power electronic controls, AFM will pave the way for building grid-friendly networked microgrids, which will significantly contribute to grid resiliency.

Currently, microgrids are typically grid-activated, meaning they are *reactive* to grid outages and abnormalities [1]. Today's microgrids are usually inverter-dominant, meaning they are highly sensitive to grid disturbances. Once grid contingencies are sensed, those inverters would be immediately tripped off or deactivated. Should there be a high penetration of networked microgrids with such hair-trigger settings, a minor fault in the grid could cause a sudden loss of a significant number of DERs and microgrids. This destabilizing impact may cascade and propagate and eventually increase the risk of a major blackout. To address this issue, the newly amended IEEE Standard 1547 [2] has recommended a wider ride-through tolerance for microgrids to ensure that microgrids inject real and reactive power after grid faults occur. The main purpose is to make microgrids support bulk power grid stability and reduce the potential excursions of frequencies and voltages [1]. However, there is another critical challenge: if NMs intend to ride through grid faults, they will inject fault currents and contribute power in an uncontrolled or uncoordinated manner, and then create disastrous impacts on the distribution grid to which they are connected. Such impacts may include (1) a high fault current level beyond the capacity of protective devices and major equipment of the grid, (2) voltage violations, (3) voltage or transient instabilities of local grid, and (4) severe power quality issues. Meanwhile, such disturbances caused in the utility grid would likely propagate back to NMs through electrical feeders (because those AC links offer synchronized interconnection and cannot block the propagation of disturbances) and plague microgrid end users. To mitigate against such detrimental impacts, utility companies and microgrid owners face installing prohibitively costly upgrades such as fault current limiters [3], remedial action schemes [4], and compensators [5]. However, even these expensive "fit and forget" solutions can hardly accommodate the fast changes in loads, microgrid plug-ins and power distribution networks.

How can we resolve this major obstacle to integrating networked microgrids in distribution grids? This book develops a potentially transformative concept called *active fault management* (AFM), which will enable *active* responses to grid abnormalities such that networked microgrids can serve as robust resiliency sources for supporting grid recovery. Instead of forcing a microgrid to change into the islanded mode, the AFM system enables a microgrid to remain in the system so that it can continue contributing power to the main grid during grid faults. This new AFM technology will significantly improve both utility grid resiliency and power supply availability for microgrid customers.

6.2   Multifunctional AFM to Enable Microgrid Survivability

Fault management for microgrids is still an open problem due to the complexity and stringent requirements for microgrids. Even for the most recent developments in microgrid technology, such as the *hybrid* AC/DC microgrid [6] and the nonsynchronous microgrid [7], leave basic fault management functions such as fault ride-through (FRT) largely unexplored [8].

So far, FRT is mainly designed for individual distributed generators [9]. While a few European codes now recommend integrating FRT in large PV plants interconnected to medium-voltage networks so that PV plants can participate in grid voltage regulation [10, 11], it is still a challenge for FRT to effectively cope with unbalanced situations and stability issues.

In this book, we focus on building the AFM for power electronic interfaced microgrids. Various remaining difficulties need to be resolved, such as the following:

- The dilemma posed by the high penetration of microgrids: how can fault ride-through be achieved while ensuring near-zero additional fault current injection to the main grid?
- The issue of microgrid stability: How can a power balance be maintained within microgrids so that hazardous power ripples and DC voltage collapses can be prevented? And, how can this be done while still resolving the aforementioned dilemma?
- How can ultrafast AFM responses be ensured for low-inertia microgrids?

Resolving these challenges will make microgrids more resilient under grid faults, which will lead to more resilient and grid-friendly networked microgrids.

The basic idea of microgrid AFM is to manage the fault current by controlling the power electronic interface (e.g., the back-to-back inverters). Besides regulating the fault current, microgrids should also provide stability support and other ancillary support [12], if necessary. This will lead to a multifunctional AFM.

For a single microgrid, the multifunctional AFM can be derived using the following procedures:

- A new philosophy is proposed to maintain the magnitude of the total fault current (sum of the fault contributions from the main grid and the microgrid) unchanged by

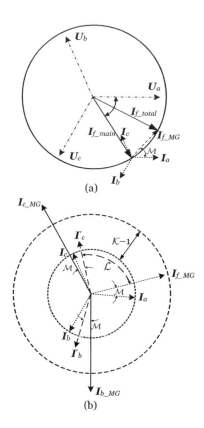

Figure 6.1 AFM vector scheme for an SLTG grid fault. (a) Shift microgrid fault current angle by $\mathcal{M}$ to limit total fault current magnitude; (b) change magnitudes and angles of phases b and c currents to eliminate power ripples, and multiply the abc currents by $\mathcal{K}$ to keep pre- and postfault power flows identical.

shifting the phase angle of the microgrid fault injection through the inverter control (see Figure 6.1 for a single-line-to-ground (SLTG) fault case).

- Unbalanced voltage sags can induce power ripples inside and outside of the microgrid. The PWM current control of the microgrid interface will generate unbalanced current injections into the grid to eliminate damaging power ripples. We can formulate the power ripples using instantaneous power theory and phase-coordinate quantities. By properly setting the angle displacements between three-phase microgrid currents, the double line-frequency power ripples can be partly or even entirely canceled out.

- The microgrid's power flow should be roughly identical before and after the fault to maintain AC subgrid stability, which can be achieved by altering the microgrid's three-phase current magnitudes. Solving all the three conditions will lead to the desirable reference currents for the microgrid inverter described by abc phase quantities, namely $I_{abc}^* = I_{abc}(x)$, where x includes local states (e.g., the subgrid's power flow, voltages, and currents), and those states of the interface buses that tie to neighboring microgrids and the main grid (e.g., magnitudes and phase angles of fault currents from the main grid, and the depth of voltage sag at the point of common coupling [PCC]).

6.3 Distributed AFM for Networked Microgrids

Multifunctional AFM is highly desirable, but its objectives may conflict with each other. Thus, finding an analytical solution to the AFM of a single microgrid can be difficult, if not impossible. Furthermore, satisfying multiple objectives may result in other issues such as an increase in the DC link voltage or in the inverter's output current beyond safety thresholds. For the previously listed reasons, it is more feasible and desirable to find an optimal solution for microgrid AFM to achieve a trade-off among multiple objectives.

As microgrid stakeholders are increasingly concerned about online security and privacy, it is unlikely that a microgrid in an NM system would give neighboring microgrids or distribution utilities the right to access its data. Rather, only a small amount of interface data would be shared with the NM community and utility company. This necessitates the design of a distributed AFM to coordinate NM fault management while protecting customer privacy. To this end, a newly developed optimization technique called Distributed and Asynchronous Surrogate Lagrangian Relaxation (DA-SLR) is introduced to enable efficient and privacy-preserving AFM for the NM system. This chapter offers an introduction to the distributed AFM, as well as its implementation and verification on a multicore PC.

6.4 Problem Formulation

Once a balanced or unbalanced fault occurs, AFM's goal is to maintain the total fault current unchanged to avoid detrimental impacts on the utility grid, to eliminate damaging power ripples for inverters in the microgrids, and to ensure that the microgrids' power flows are identical before and after fault to avoid loss of loads and maintain NM stability. We formulate AFM for NMs as a multiobjective nonlinear optimization problem. The solution for this multiobjective optimization problem should be efficient so that it can be implemented online. To ensure scalability, the AFM formulation and solution should be *reconfigurable* and support *plug-and-play* [13–15] to adapt to ever-evolving grid conditions and levels of microgrid participation.

We formulate AFM in rectangular coordinates to facilitate more efficient optimization solutions. The AFM problem has two objectives to minimize:

1. J_1 – The total contribution from NMs in increasing the main grid's fault current magnitude
2. J_2 – double-line-frequency ripples in the instantaneous power that microgrids inject to the main grid

The AFM for NMs can then be formulated as follows:

$$\text{min} \qquad \alpha J_1 + (1 - \alpha)J_2, \quad \alpha \in [0, 1] \tag{6.1}$$

s.t.

$$\sum_j [\text{Re}(\mathbf{U}_{i,j})\text{Re}(\mathbf{I}_{i,j}) + \text{Im}(\mathbf{U}_{i,j})\text{Im}(\mathbf{I}_{i,j})] = P_i \tag{6.2}$$

$$\sum_j \mathbf{I}_{i,j} = \mathbf{0} \tag{6.3}$$

$$[\text{Re}(\mathbf{I}_{i,j})]^2 + [\text{Im}(\mathbf{I}_{i,j})]^2 \le (I_i^R)^2 \tag{6.4}$$

$$\left[\text{Re}\left(\sum_{i=1}^N \mathbf{I}_{i,j} \right) \right]^2 + \left[\text{Im}\left(\sum_{i=1}^N \mathbf{I}_{i,j} \right) \right]^2 \le (I^{R,M})^2, \tag{6.5}$$

where

$$\begin{cases} J_1 \equiv \frac{1}{D} \sum_\varphi \left| \frac{[\text{Re}(\mathbf{I}_\varphi^M + \mathbf{I}_\varphi^m)]^2 + [\text{Im}(\mathbf{I}_\varphi^M + \mathbf{I}_\varphi^m)]^2}{[\text{Re}(\mathbf{I}_\varphi^M)]^2 + [\text{Im}(\mathbf{I}_\varphi^M)]^2} - 1 \right| \\[2mm] J_2 \equiv \frac{1}{N} \sum_{i=1}^N \frac{P_i^r}{(P_i)^2} \\[2mm] \mathbf{I}_\varphi^m \equiv \sum_{i=1}^N (\mathbf{S}_i \mathbf{I}_{i,f}) \\[2mm] P_i^r \equiv \{\sum_j [\text{Re}(\mathbf{U}_{i,j})\text{Re}(\mathbf{I}_{i,j}) - \text{Im}(\mathbf{U}_{i,j})\text{Im}(\mathbf{I}_{i,j})]\}^2 \\[2mm] \qquad + \{\sum_j [\text{Im}(\mathbf{U}_{i,j})\text{Re}(\mathbf{I}_{i,j}) + \text{Re}(\mathbf{U}_{i,j})\text{Im}(\mathbf{I}_{i,j})]\}^2 \end{cases} \tag{6.6}$$

$$(i = 1, \dots, N; j = a, b, c; \varphi \in \mathcal{P}(\{a, b, c\})).$$

Solving the optimization problem will find microgrid i's reference currents $\text{Re}(\mathbf{I}_{i,j})$ and $\text{Im}(\mathbf{I}_{i,j})(j = a, b, c)$ that fulfill the AFM requirements. Here $\mathbf{U}$ and $\mathbf{I}$ represent voltage and current phasors, respectively. N is the total number of networked microgrids. Superscripts M and m are used to denote variables of the main grid and all the networked microgrids, respectively. Superscript R indicates the equipment's emergency rating (e.g., Rate B for a utility line). Subscript i represents microgrid i. Subscript j denotes all three phases $\{a, b, c\}$ while φ represents the faulted phases represented by a powerset [16] of $\{a, b, c\}$. For instance, if a double-line-to-ground fault occurs between phases a and b, then φ is $\{a, b\}$.

α is the weight factor that represents the trade-off between J_1 and J_2. $\mathbf{S}_i$ is a complex number that represents the fault current's phase jump caused by transformer(s) winding connections between microgrid i and the fault location. P_i is the active power injection from microgrid i to the main grid. I_i^R is the emergency rating of microgrid i's interface, and $I^{R,M}$ is the emergency rating of the feeder connecting the main grid and the NMs.

In objective J_1, $[\text{Re}(\mathbf{I}_\varphi^M)]^2 + [\text{Im}(\mathbf{I}_\varphi^M)]^2$ and $[\text{Re}(\mathbf{I}_\varphi^m)]^2 + [\text{Im}(\mathbf{I}_\varphi^m)]^2$ are fault currents (actually their squares for computational convenience) from the main grid and NMs, respectively. $[\text{Re}(\mathbf{I}_\varphi^M + \mathbf{I}_\varphi^m)]^2 + [\text{Im}(\mathbf{I}_\varphi^M + \mathbf{I}_\varphi^m)]^2$ is the total fault currents (the sum of the main grid and the NMs' fault contributions). By minimizing J_1, the total fault current magnitude will be kept unchanged with any numbers of networked microgrids tied to the main grid. In objective J_2, P_i^r represents the amplitude of the power ripples flowing through microgrid i's grid interface (again, actually their squares). In the aggregated objective (6.1), we introduce a scaling factor D to scale J_1 to the same order of J_2 to improve the accuracy of the optimization results.

Among the four constraints, (6.2) is to keep the power balanced before and after faults to maintain the voltage stability of the grid-tie inverter and the dynamic stability of the NMs; (6.3) is to make the AFM work for an ungrounded grid; (6.4) and (6.5) are to ensure the fault currents flowing through microgrid i and the main grid remain within their emergency ratings so that no damage occurs during the fault ride-through.

6.5 A Distributed Solution to AFM

To efficiently solve the AFM problem while preserving customer privacy, DA-SLR is used to solve the aforementioned optimization problem in Section 6.4. DA-SLR has been mathematically proven to be efficient with guaranteed convergence [17]. The multicore implementation of the distributed optimization solution for AFM will be introduced, and the method is validated on a six-microgrid test case.

6.5.1 Basics of Lagrangian Relaxation

This section gives a brief and incomplete discussion of Lagrangian relaxation to clarify the DA-SLR solution for AFM in NMs. Readers are encouraged to read [18] for mathematical details and rigorous treatment of these subjects.

Lagrangian relaxation (LR) is an approach to solving optimization problems with constraints. In unconstrained problems, the necessary condition for a minimizer is $\nabla f = 0$. Lagrange multipliers and Lagrangian relaxation are introduced so that similar necessary conditions exist for constrained problems.

In problem (6.7), if there are no constraints (6.8)–(6.11), $\nabla f(x) = 0$ is the necessary condition for the vector x^* to be the minimizer. However, because of constraints (6.8)–(6.11), a minimizer x^* may not satisfy $\nabla f(x) = 0$.

$$\min \quad f(x) \tag{6.7}$$

$$\text{s.t.} \quad h_1(x) = 0 \tag{6.8}$$

$$h_2(x) = 0 \tag{6.9}$$

$$g_1(x) \leq 0 \tag{6.10}$$

$$g_2(x) \leq 0, \tag{6.11}$$

where f, h_i, g_i are real-valued functions on $\mathbb{R}^n$, and $x \in \mathbb{R}^n$.

Lagrangian relaxation is introduced to tackle the problem by forming the following Lagrangian function:

$$L(x, \lambda, \mu) = f(x) + \lambda_1 h_1(x) + \lambda_2 h_2(x) + \mu_1 g_1(x) + \mu_2 g_2(x), \tag{6.12}$$

where $\lambda = [\lambda_1, \lambda_2]^T$ are Lagrange multipliers for equality constraints, and $\mu = [\mu_1, \mu_2]^T$ are Lagrange multipliers for inequality constraints. After relaxation, the necessary condition for a minimizer of problem (6.7)–(6.11) becomes $\nabla L_x(x, \lambda, \mu) = 0$.

For a few special types of problems, LR can be used to decompose the problem into several subproblems. The following optimization problem is taken as an example:

$$\min \qquad f_1(x_1) + f_2(x_2) + f_3(x_3) \tag{6.13}$$

$$\text{s.t.} \qquad h_{11}(x_1) + h_{21}(x_2) + h_{31}(x_3) = 0 \tag{6.14}$$

$$h_{12}(x_1) + h_{22}(x_2) + h_{32}(x_3) = 0, \tag{6.15}$$

where f_i, h_{ij} are real-valued functions. x_1, x_2, and x_3 are three decision variables.

The corresponding Lagrangian function is

$$L(x, \lambda) = f_1(x_1) + f_2(x_2) + f_3(x_3) + \lambda_1[h_{11}(x_1) + h_{21}(x_2) + h_{31}(x_3)]$$
$$+ \lambda_2[h_{12}(x_1) + h_{22}(x_2) + h_{32}(x_3)]. \tag{6.16}$$

When the Lagrangian method is used to solve the problem (6.13)–(6.15), the updating takes the following form,

$$\begin{cases} x_1^{k+1} = x_1^k - \epsilon^k \frac{dL(x^k, \lambda^k)}{dx_1}\Big|_{x_1^k} = x_1^k - \epsilon^k \frac{d[f_1(x_1) + \lambda_1^k h_{11}(x_1) + \lambda_2^k h_{12}(x_1)]}{dx_1}\Big|_{x_1^k} \\ x_2^{k+1} = x_2^k - \epsilon^k \frac{dL(x^k, \lambda^k)}{dx_2}\Big|_{x_2^k} = x_2^k - \epsilon^k \frac{d[f_2(x_2) + \lambda_1^k h_{21}(x_2) + \lambda_2^k h_{22}(x_2)]}{dx_2}\Big|_{x_2^k} \\ x_3^{k+1} = x_3^k - \epsilon^k \frac{dL(x^k, \lambda^k)}{dx_3}\Big|_{x_3^k} - x_3^k - \epsilon^k \frac{d[f_3(x_3) + \lambda_1^k h_{31}(x_3) + \lambda_2^k h_{32}(x_2)]}{dx_3}\Big|_{x_3^k} \end{cases} \tag{6.17}$$

$$\begin{cases} \lambda_1^{k+1} = \lambda_1^k + \epsilon^k [h_{11}(x_1^k) + h_{21}(x_2^k) + h_{31}(x_3^k) \\ \lambda_2^{k+1} = \lambda_2^k + \epsilon^k [h_{12}(x_1^k) + h_{22}(x_2^k) + h_{32}(x_3^k)], \end{cases} \tag{6.18}$$

where ϵ^k is step size. As we can see in (6.17), the updating formulation for x_i only contains x_i without having other decision variables. For example, x_1's updating formulation only has x_1 and does not have x_2 and x_3. In this way, the problem (6.13)–(6.15) can be decomposed into three subproblems, whose Lagrangian functions are expressed in the following three equations:

$$L(x_1, \lambda) = f_1(x_1) + \lambda_1[h_{11}(x_1) + h_{21}(x_2) + h_{31}(x_3)] + \lambda_2[h_{12}(x_1) + h_{22}(x_2) + h_{32}(x_3)] \tag{6.19}$$

$$L(x_2, \lambda) = f_2(x_2) + \lambda_1[h_{11}(x_1) + h_{21}(x_2) + h_{31}(x_3)] + \lambda_2[h_{12}(x_1) + h_{22}(x_2) + h_{32}(x_3)] \tag{6.20}$$

$$L(x_3, \lambda) = f_3(x_3) + \lambda_1[h_{11}(x_1) + h_{21}(x_2) + h_{31}(x_3)] + \lambda_2[h_{12}(x_1) + h_{22}(x_2) + h_{32}(x_3)]. \tag{6.21}$$

After decomposition, problems (6.19)–(6.21) can be solved with three different computation units, realizing parallel computation. Another coordinating computation unit is needed to update the Lagrange multipliers based on (6.18). During each iteration, the ith computation unit sends its value x_i to the central unit, and receives λ from the coordination unit.

Thus, the key idea of LR has two parts: decomposition and coordination. In decomposition, by introducing the Lagrange multipliers λ, the coupled system constraints are relaxed. The relaxing problem results in smaller subproblems, which are easier to solve than the original problem.

The major difference between Surrogate Lagrangian Relaxation (SLR) and LR is that in LR, all the subproblems must be solved to update the Lagrange multipliers, while in SLR only one subproblem needs to be solved to update the Lagrange multipliers. Actually, the problem formulation for SLR and LR are the same: by introducing the Lagrange multipliers λ, centralized formulation (6.13)–(6.15) is decomposed into three subproblems (6.19)–(6.21).

In LR, the coordination unit updates λ based on (6.18) after all three subproblems (6.19)–(6.21) are solved. In SLR, the coordination unit updates λ based on (6.18) after any of the three subproblems (6.19)–(6.21) is solved. SLR needs a prudent selection of the step size for multipliers updating, and the proof of its convergence is nontrivial [19]. Based on SLR, our distributed AFM is solved through a two-level iterative approach. The low level, decomposition, is responsible for solving individual subproblems, and the upper level, coordination, is responsible for updating the Lagrange multipliers.

6.5.2 Solving AFM Using Distributed and Asynchronous SLR

The AFM problem modeled in (6.1)–(6.5) is largely separable except for constraint (6.5), which involves fault current contributions from all the microgrids. This constraint can be relaxed with the Lagrange multipliers λ[17, 19]. The relaxed problem is further decomposed into N subproblems, each of which is the AFM of a single microgrid i'. For a better understanding, microgrid i''s AFM in the kth iteration (see Algorithm 1) is expressed as follows:

$$\min \quad \alpha J_{i',1} + (1 - \alpha)J_{i',2} + \lambda^T g, \quad \alpha \in [0,1] \tag{6.22}$$

s.t.

$$\sum_j [\mathrm{Re}(\mathbf{U}_{i',j})\mathrm{Re}(\mathbf{I}_{i',j}) + \mathrm{Im}(\mathbf{U}_{i',j})\mathrm{Im}(\mathbf{I}_{i',j})] = P_{i'} \tag{6.23}$$

$$\sum_j \mathbf{I}_{i',j} = \mathbf{0} \tag{6.24}$$

$$(\mathrm{Re}(\mathbf{I}_{i',j}))^2 + (\mathrm{Im}(\mathbf{I}_{i',j}))^2 \le (I_{i'}^R)^2, \tag{6.25}$$

where

$$\begin{cases} J_{i',1} \equiv \dfrac{1}{D}\sum_f \left| \dfrac{[\mathrm{Re}(\mathbf{I}_f^M + \mathbf{I}_f^m)]^2 + [\mathrm{Im}(\mathbf{I}_\varphi^M + \mathbf{I}_\varphi^m)]^2}{[\mathrm{Re}(\mathbf{I}_\varphi^M)]^2 + [\mathrm{Im}(\mathbf{I}_\varphi^M)]^2} - 1 \right| \\[4mm] \mathbf{I}_\varphi^m \equiv \displaystyle\sum_{i=1, i\ne i'}^N (\mathbf{S}_i \mathbf{I}_{i,\varphi}^{k-1}) + \mathbf{S}_{i'}\mathbf{I}_{i',\varphi} \end{cases} \tag{6.26}$$

$$g_j \equiv \left[\text{Re} \left(\sum_{i=1, i \neq i'}^{N} \mathbf{I}_{i,j}^{k-1} + \mathbf{I}_{i',j} \right) \right]^2 + \left[\text{Im} \left(\sum_{i=1, i \neq i'}^{N} \mathbf{I}_{i,j}^{k-1} + \mathbf{I}_{i',j} \right) \right]^2 - (I^{R,M})^2$$

(6.27)

$$(j = a, b, c; \varphi \in \mathcal{P}(\{a, b, c\})).$$

Here the optimization is to find current references $\text{Re}(\mathbf{I}_{i',j})$ and $\text{Im}(\mathbf{I}_{i',j})(j = a, b, c)$ for microgrid i'. $\boldsymbol{\lambda} = [\lambda_a, \lambda_b, \lambda_c]$ and $\boldsymbol{g} = [g_a, g_b, g_c]$ represent the Lagrange multipliers and violations due to the relaxation of constraint (6.5), respectively.

The subproblem expressed by (6.22)–(6.27) is solved with sequential quadratic programming (SQP). The basic idea of SQP is to approximate the original problem with a quadratic programming subproblem at the solution of iteration k', $I_{i'}^{k'}$. Then, $I_{i'}^{k'+1}$, the solution of iteration $k'+1$, is achieved by solving the approximate quadratic subproblem [20]. This process is iterated until the solution converges. The approximation is achieved by linearization at $I_{i'}^{k'}$. The quadratic subproblem with quadratic objective functions and linear constraints can be efficiently solved.

Advantages of SQP include its ability to solve problems with nonlinear constraints and the fact that it does not require feasible points at the initial points or at each of the iterations. Finding feasible points for nonlinear programming (NLP) may be as hard as solving NLP.

With SQP, the subproblem of the distributed AFM (6.22)–(6.27) is linearized as the following quadratic subproblem:

$$\min \quad \nabla L(I_{i'}^{k'}, \boldsymbol{u}^{k'}, \boldsymbol{v}^{k'}) d\mathbf{I} + \frac{1}{2} (d\mathbf{I})^T \mathbf{B}^{k'} d\mathbf{I}$$

(6.28)

s.t.

$$\nabla \boldsymbol{h}(I_{i'}^{k'})^T d\mathbf{I} + \boldsymbol{h}(I_{i'}^{k'}) = 0$$

(6.29)

$$\nabla \boldsymbol{w}(I_{i'}^{k'})^T d\mathbf{I} + \boldsymbol{w}(I_{i'}^{k'}) \leq 0,$$

(6.30)

where

$$L(I_{i'}^{k'}, \boldsymbol{u}^{k'}, \boldsymbol{v}^{k'}) = \alpha J_{i',1} + (1 - \alpha) J_{i',2} + \boldsymbol{\lambda}^T \boldsymbol{g} + \boldsymbol{h}^T \boldsymbol{u}^{k'} + \boldsymbol{w}^T \boldsymbol{v}^{k'} \quad \alpha \in [0, 1]$$

(6.31)

$$\begin{cases} h_1 = \sum_j [\text{Re}(\mathbf{U}_{i',j}) \text{Re}(\mathbf{I}_{i',j}) + \text{Im}(\mathbf{U}_{i',j}) \text{Im}(\mathbf{I}_{i',j})] - P_{i'} \\ h_2 = \text{Re}(\sum_j \mathbf{I}_{i',j}) \\ h_3 = \text{Im}(\sum_j \mathbf{I}_{i',j}) \\ w = (\text{Re}(\mathbf{I}_{i',j}))^2 + (\text{Im}(\mathbf{I}_{i',j}))^2 - (I_{i'}^R)^2. \end{cases} \quad j = a, b, c \quad (6.32)$$

Subproblem solutions are coordinated by the coordinator, which can be implemented in the NM management system, the tertiary controller, or the distribution management system by updating the Lagrange multipliers

$$\boldsymbol{\lambda}^{k+1} = \max(\boldsymbol{\lambda}^k + \epsilon^k \boldsymbol{g}^k, \mathbf{0}),$$

(6.33)

where k is the iteration number; ϵ^k is the step size; and g^k is given by (6.27). The coordinator updates λ immediately after one microgrid finishes its subproblem optimization (6.22)–(6.27) without waiting for all the microgrids to finish solving the subproblems. This is why the approach is called distributed and asynchronous SLR (DA-SLR).

The convergence of DA-SLR has been proved [17] for separable problems in which each subproblem is independent from the solutions of the other subproblems. The distributed AFM problem is not fully separable in that the quadratic terms within (6.26) and (6.27) include current phasors of all microgrids. Fortunately, when the subproblem modeled by (6.22)–(6.27) is solved subject to the surrogate optimality condition, the resulting multiplier-updating direction forms acute angles with directions toward the optimum. Furthermore, because subproblems (6.22)–(6.27) so far do not involve discrete variables and each of them is a medium-sized problem, this greatly lowers the risk if divergence and convergence is guaranteed when the step size is properly selected following the guide in [17].

6.5.3 Implementation of Distributed AFM on Multiple Computation Cores

Even though there is a plethora of literature for distributed optimization, there is a lack of real implementation of distributed optimization in distributed computing resources. Rather, most of the literature on this subject has adopted a pseudodistributed approach [21, 22] where all the subproblems are solved in one computation core with a predefined sequence of the subproblems' solutions. We choose to implement the distributed AFM algorithm through MATLAB's Parallel Computing Toolbox [23] in a distributed multicore environment. Each subproblem is assigned to a different core, and the calculation sequence is dynamically determined by the computer cores' processing capabilities instead of being predefined. This will gain real-life insights for the distributed and asynchronous algorithm for the distributed AFM and help test the capability of the distributed AFM for field applications.

Figure 6.2 shows the schematic for implementing the distributed AFM in multiple cores. Before each computation, core i' receives the newest value of the Lagrange

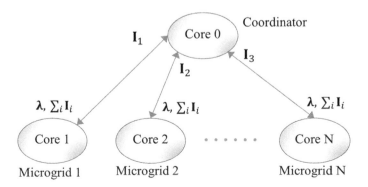

Figure 6.2 Schematic for the distributed and asynchronous implementation of distributed AFM within one CPU of multiple cores. $\mathbf{I}_i$ is microgrid i's currents.

multipliers λ and the sum of the NMs' fault currents $\sum_i \mathbf{I}_i$ from the coordinator. After computation finishes, core i' sends microgrid i''s AFM results $\mathbf{I}_{i'}$ back to the coordinator. The coordinator, running on one individual core, updates λ and $\sum_i \mathbf{I}_i$ immediately upon receiving the updated outputs from the first microgrid that finishes its AFM calculation (assuming it is microgrid i') without waiting for results from the other microgrids/cores. The overall computation process of the distributed AFM is summarized in Algorithm 1, where it ends after λ converges. The SQP calculation is summarized in Algorithm 2.

As shown in Figure 6.2 and Algorithm 1, the computation core for each microgrid, which in the real world represents the local computing resource within the microgrid or cloud computing resource, updates asynchronously without waiting until other microgrids finish their computation and updating processes. Therefore, the distributed AFM fully supports the microgrids' plug-and-play. After the connection or disconnection of certain microgrids, no modification in the AFM algorithm is required for the rest of the microgrids. The only slight change is the computation of $\sum_i \mathbf{I}_i$ in the

---

Algorithm 1 Distributed active fault management for networked microgrids

1: **Result:** N microgrids' output currents $\mathbf{I}_i (i = 1, \dots, N)$
2: **Initialization:** N microgrids' currents $\mathbf{I}_i^0 (i = 1, \dots, N)$, Lagrange multipliers λ^0, updating step size ϵ^0, termination criterion σ
3: iteration $k \leftarrow 1$
4: **while** $\| \lambda^k - \lambda^{k-1} \| < \sigma$ **do**
5: N microgrids optimize distributedly with SQP based on (6.22) through (6.27)
6: Microgrid i' finishes its optimization, outputs its currents $\mathbf{I}_{i'}^k$
7: The coordinator updates λ^k asynchronously with $\mathbf{I}_{i'}^k$ and $\mathbf{I}_i^{k-1}$ ($i = 1, \dots, i' - 1, i' + 1, \dots, N$) based on (6.33)
8: iteration $k \leftarrow k + 1$
9: **end while**

Algorithm 2 Sequential Quadratic Programming (SQP)

1: **Results:** microgrid i''s output currents $\mathbf{I}_{i'}$
2: **Initialization:** microgrid i''s currents $\mathbf{I}_{i'}^0$; a finite difference approximation $\mathbf{B}^0$ of the actual Hessian $\mathbf{H}L$
3: iteration $k' \leftarrow 0$
4: **while** $\| d\mathbf{I} \| < \sigma$
5: Get the quadratic subproblem (6.28)–(6.32) by linearation at $\mathbf{I}_{i'}^{k'}$; solve it to get $d\mathbf{I}$
6: Update: $\mathbf{I}_{i'}^{k'+1} = \mathbf{I}_{i'}^{k'} + d\mathbf{I}$
7: Update: $\mathbf{B}^{k'+1} = \mathbf{B}^{k'} + \frac{1}{d\mathbf{I}^T d\mathbf{I}}[(y - \mathbf{B}^{k'} d\mathbf{I})d\mathbf{I}^T + d\mathbf{I}(y - \mathbf{B}^{k'} d\mathbf{I})^T] - \frac{(y - \mathbf{B}^{k'} d\mathbf{I})^T}{(d\mathbf{I}^T d\mathbf{I})^2} d\mathbf{I} d\mathbf{I}^T$, with $y = \nabla L(I_{i'}^{k'+1}, u^{k'+1}, v^{k'+1}) - \nabla L(I_{i'}^{k'}, u^{k'}, v^{k'})$
8: iteration $k' \leftarrow k' + 1$
9: **end while**

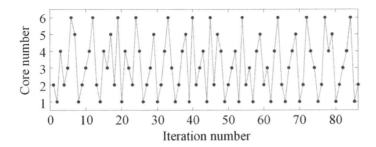

Figure 6.3 Calculation sequence of distributed AFM with six microgrids for one optimization. Each core solves AFM for one microgrid.

coordinator's algorithm. The distributed AFMs' distributed and asynchronous nature can better facilitate the NMs' future development, where heterogeneous microgrids of different owners and computation capacities will be connected together to further the microgrids' benefits. Figure 6.3 shows the calculation sequence of distributed AFM with six microgrids for one optimization.

6.6   Testing and Validation

Extensive case studies on a six-microgrid NM system are used to test and validate the effectiveness of the distributed AFM. The parameters of the NM system are summarized in Table 6.1. Three types of grounded faults are applied to the main grid feeder to test the performance of the distributed AFM scheme. A weight factor $\alpha = 0.95$ is adopted for (6.1) and (6.22).

To validate the performance, the distributed AFM is compared with a centralized AFM solution and a naive ride-through method:

- Comparison with the centralized approach: The centralized AFM shares the same formulation (6.1)–(6.6) with the distributed AFM. In contrast to the distributed AFM, the centralized approach has access to information from all of the microgrids in the NM system. The centralized AFM is solved directly by SQP without requiring iterative updating from individual microgrids and thus does not suffer from potential divergence issues. The comparison between the centralized and distributed approaches aims to verify the convergence performance of the distributed AFM.
- Comparison with a "naive" ride-through approach: A naive ride-through method only ensures NM ride-through and manages power balances in microgrids (this implies maintaining the DC link voltages of grid-tie inverters and limiting the microgrids' output currents below safety limits) regardless of mitigating power ripples and limiting the total fault currents. This comparison is to demonstrate the advantages of the distributed AFM in enabling a resilient NM system and utility grid.

Table 6.1 NM system parameters including PI control parameters.

Microgrids (MGs)	Capacity (kW)	Power delivery to grid (kW)	DC voltages control of distributed AFM	Current angle control of distributed AFM	Amplitude control of the simple ride-through method	DC voltages control under normal conditions	Reactive power control under normal conditions
MG 1	852	213					
MG 2	624	156				2.50 (P)	
MG 3	884	221	0.014 (P)	0.006 (P)	0.014 (P)	37.50 (I)	$2.5 \cdot 10^{-4}$ (P)
			0.450 (I)	0.600 (I)	0.450 (I)	———	0.040 (I)
MG 4	1208	302				3.333 (P)	
MG 5	1524	381				50.00 (I)	
MG 6	2112	528					

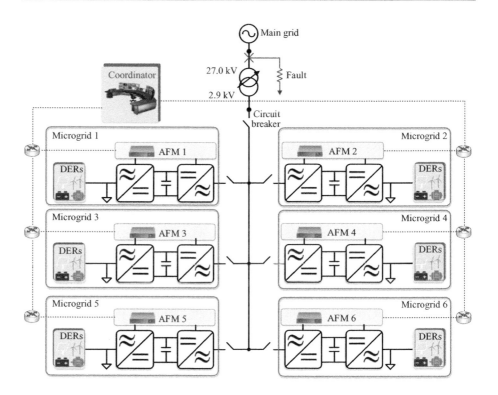

Figure 6.4 Distributed AFM for networked microgrids.

The test system is illustrated in Figure 6.4, where each microgrid in the NM system connects to the main feeder through a back-to-back inverter. Before faults happen, the active power delivered to the main grid by Microgrid 1–Microgrid 6 are 213 kW, 156 kW, 221 kW, 302 kW, 381 kW, and 528 kW, respectively, and there are no reactive power injections from microgrids to the main grid. The basic parameters of the NM system can be found in Table 6.1. For the three types of grounded faults, the distributed AFM demonstrates excellent convergence, achieves the same results as the centralized AFM, and performs consistently better than the naive ride-through method. The distributed AFM is expected to prevail over the centralized approach in that it preserves privacy, supports plug-and-play, is computationally scalable and can be easily implemented in the future software-defined networked microgrid platforms.

6.6.1   Single-Line-to-Ground Fault

In this case, a single-line-to-ground fault occurs at 0.6 s and is cleared at 0.9 s. The phase a fault causes a voltage sag of residual voltage with a value of 0.50 p.u. Figure 6.5 shows the NMs' power outputs and phase a fault currents flowing into the fault point with the distributed AFM. Figure 6.6 shows NMs' terminal voltages, currents flowing into the grid, and DC link voltages with the distributed AFM. As shown in Figure 6.5, the fault current contribution from the main grid and the total fault current are managed to be at the same amplitude of 154.31 A, meaning NMs' contributions to fault currents is 0%. Power ripples of six microgrids are reduced to 0.47%, 1.60%, 2.04%, 1.82%, 0.53%, and 0.47%, respectively, with an average value of 1.15%.

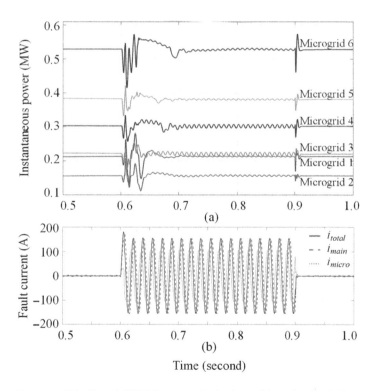

Figure 6.5 Distributed AFM for networked microgrids under single-line-to-ground fault: (a) Microgrid power outputs; (b) phase a currents at fault location. i_{total} = total fault current to the ground; i_{main} = fault current contribution from the main grid; i_{micro} = fault current contribution from NMs.

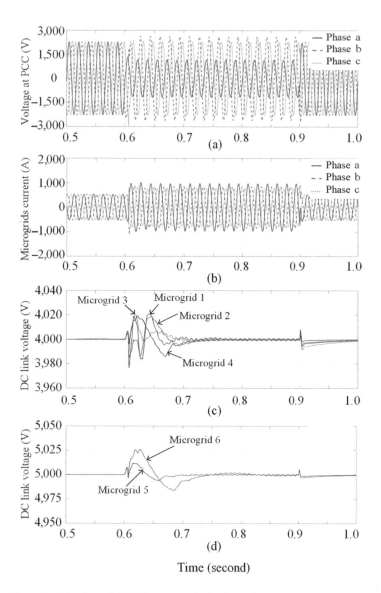

Figure 6.6 Distributed AFM for networked microgrids under single-line-to-ground fault: (a) NMs terminal voltages at PCC; (b) NMs currents flowing to the main grid; and (c) and (d) NMs DC link voltages.

Figure 6.7 shows the results obtained from the centralized AFM and the naive ride-through method under single-line-to-ground fault. For the centralized AFM (Figure 6.7a,b), the fault current contributions are also managed down to 0.00% and power ripples through the six microgrid inverters are 1.17%, 1.60%, 0.79%, 0.83%, 0.66%, and 0.38%, respectively, with an average value of 0.91%. It can be seen that the results from the distributed AFM (0.00% fault current contributions and 1.15% power ripples on average) are nearly identical to those from the centralized

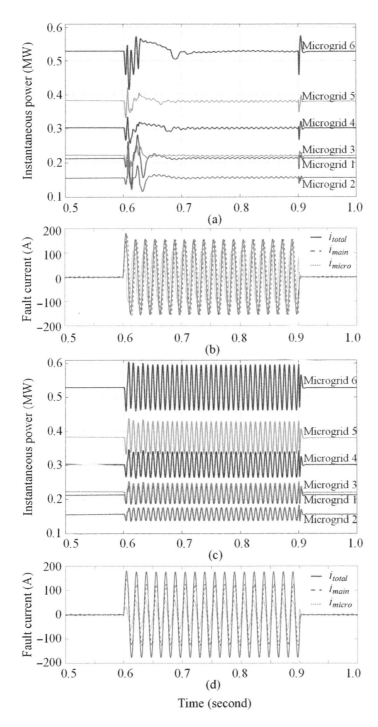

Figure 6.7 Centralized AFM for networked microgrids under single-line-to-ground fault:
(a) Microgrid power outputs; (b) phase *a* currents at fault location. Naive ride-through method
for networked microgrids under single-line-to-ground fault: (c) Microgrid power outputs;
(d) phase *a* currents at fault location.

Table 6.2 Performances of fault management methods under single-line-to-ground fault.

	Distributed AFM		A naive ride-through method		Centralized AFM	
	Fault contributions	Power ripples	Fault contributions	Power ripples	Fault contributions	Power ripples
MG 1		0.47%		12.91%		1.17%
MG 2		1.60%		11.86%		1.60%
MG 3	0.00%	2.04%	14.74%	11.31%	0.00%	0.79%
MG 4		1.82%		13.25%		0.83%
MG 5		0.53%		13.12%		0.66%
MG 6		0.47%		13.26%		0.38%
Average	0.00%	1.15%	14.74%	12.62%	0.00%	0.91%

AFM. For the naive ride-through method, however, the fault current contributions are 14.74% (177.05/154.31) and power ripples in the six microgrids are 12.91%, 11.86%, 11.31%, 13.25%, 13.12%, and 13.26%, respectively, with an average value of 12.62% (see Figure 6.7c,d). This means that the naive ride-through approach performs poorly and may cause detrimental impacts when the networked microgrids scale up. Table 6.2 summarizes the performances of the three fault management methods under single-line-to-ground fault.

6.6.2   Double-Line-to-Ground Fault

In this case, a double-line-to-ground fault is applied on phase a and phase b at 0.6 s and is cleared at 0.9 s with a residual voltage of 0.50 p.u.

Figure 6.8 illustrates the NMs' power outputs and phases a and b fault currents flowing into the fault point with distributed AFM. Figure 6.9 shows the NMs terminal voltages, currents flowing into the grid, and DC link voltages with distributed AFM. As shown in Figure 6.8, fault currents from the main grid and the total fault currents are managed to be at the same amplitude of 171.92 A for phase a and 201.43 A for phase b, meaning NMs' contributions to fault currents are reduced to 0.00% for faulted phases. Power ripples of six microgrids are reduced to 15.26%, 16.67%, 14.71%, 10.10%, 10.89%, and 9.66%, respectively, with an average value of 12.88%.

Figure 6.10 shows the results obtained from the centralized AFM and the naive ride-through method under a double-line-to-ground fault. For the centralized AFM (Figure 6.10a,b), the fault current contributions are also managed down to 0.00% for faulted phases, and power ripples through the six microgrids inverters are 15.26%, 21.47%, 19.23%, 13.58%, 16.14%, and 13.45%, respectively, with an average value of 16.52%, It can be seen that the results from the distributed AFM (0.00% for current contributions and 12.88% for power ripples) are nearly identical to those from the centralized AFM. For the naive ride-through method, however, the fault current contributions are 26.35% (217.22/171.92) for phase a and 16.93% (235.54/201.43) for phase b, and power ripples in the six microgrids are 13.38%, 13.78%, 13.58%,

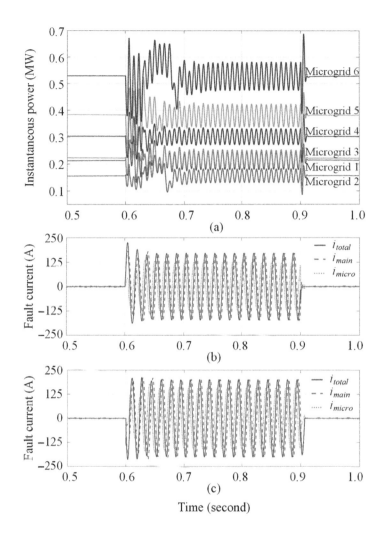

Figure 6.8 Distributed AFM for networked microgrids under double-line-to-ground fault: (a) Microgrid power outputs; (b) phase a currents at fault location; (c) phase b currents at fault location. i_{total} = total fault current to the ground; i_{main} = fault current contribution from the main grid; i_{micro} = fault current contribution from NMs.

13.41%, 13.12%, and 13.45%, respectively, with an average value of 13.45% (see Figure 6.10c,d). This means that the naive ride-through approach performs poorly and may cause detrimental impacts when the networked microgrids scale up. Table 6.3 summarizes the performances of the fault management methods for double-phase-to-ground fault.

6.6.3 Three-Phase-to-Ground Fault

In this case, a three-phase-to-ground fault is applied at 0.6 s and is cleared at 0.9 s with a residual voltage of 0.87 p.u. Three-phase fault is the most severe contingency for NMs. As the fault is symmetrical, the power ripples are zero.

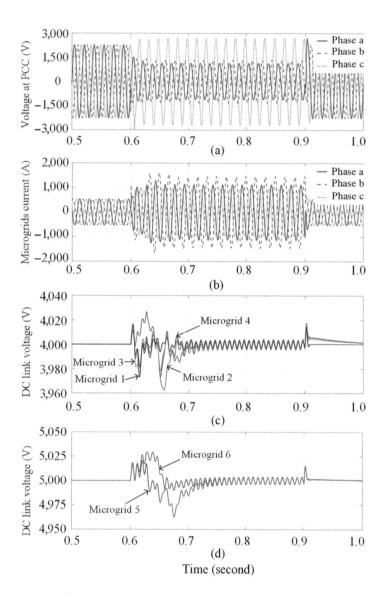

Figure 6.9 Distributed AFM for networked microgrids under double-line-to-ground fault:
(a) NMs terminal voltages at PCC; (b) NMs currents flowing to the main grid; and (c) through
(d) NMs DC link voltages.

Figure 6.11 illustrates the phase *a* fault currents flowing into the fault location with
distributed AFM. Figure 6.12 illustrates the NMs' terminal voltages, currents flowing
into the grid, and DC link voltages with distributed AFM. As shown in Figure 6.11a,
the fault currents from the main grid and the total fault currents are managed to
be 64.96 *A* and 77.57 *A*, respectively, making the NMs' fault current contributions
at 19.41%.

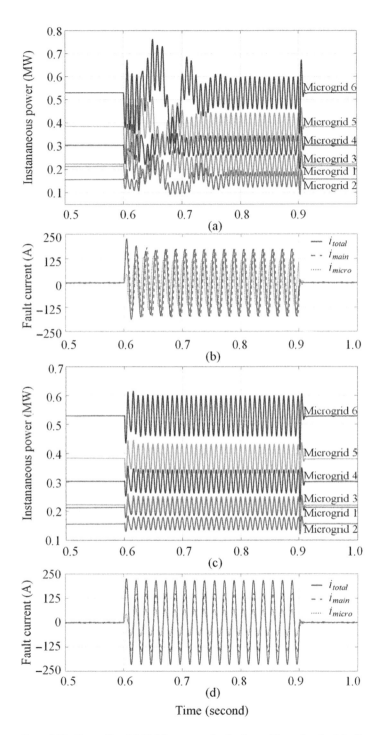

Figure 6.10 Centralized AFM for networked microgrids under double-line-to-ground fault:
(a) Microgrid power outputs; (b) phase *a* currents at fault location. Naive ride-through method
for networked microgrids under double-line-to-ground fault: (c) Microgrid power outputs;
(d) phase *a* currents at fault location.

Table 6.3 Performances of fault management methods under double-line-to-ground fault.

Microgrids (MGs)	Distributed AFM		Naive ride-through method		Centralized AFM	
	Fault contributions	Power ripples	Fault contributions	Power ripples	Fault contributions	Power ripples
MG 1		15.26%		13.38%		15.26%
MG 2	0.00%	16.67%	26.35%	13.78%	0.00%	21.47%
MG 3	(phase a);	14.71%	(phase a);	13.58%	(phase a);	19.23%
MG 4	0.00%	10.10%	16.93%	13.41%	0.00%	13.58%
MG 5	(phase b)	10.89%	(phase b)	13.12%	(phase b)	16.14%
MG 6		9.66%		13.45%		13.45%
Average	0.00%; 0.00%	12.88%	26.35%; 16.93%	13.45%	0.00%; 0.00%	16.52%

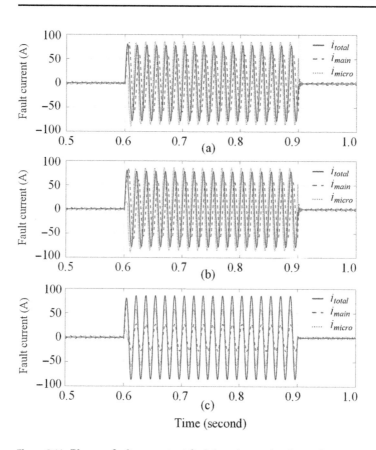

Figure 6.11 Phase a fault currents at fault location under three-phase-to-ground fault: (a) distributed AFM, (b) centralized AFM, and (c) naive ride-through method. i_{total} = total fault current to the ground; i_{main} = fault current contribution from the main grid; i_{micro} = fault current contribution from NMs.

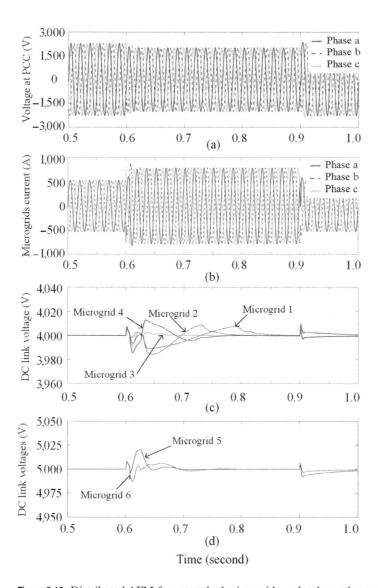

Figure 6.12 Distributed AFM for networked microgrids under three-phase-to-ground fault: (a) NMs terminal voltages at PCC; (b) NMs currents flowing to the main grid; and (c) through (d) NMs' DC link voltages.

Figure 6.11b,c show results with the centralized AFM and the naive ride-through method for three-phase-to-ground fault, respectively. For the centralized AFM, the fault current contributions are 20.03%, which is close to the 19.41% of the distributed AFM. For the naive ride-through method, the fault current contributions are 32.05% (85.78/64.96), which is worse than the 19.41% of the distributed AFM. Table 6.4 summarized the performance of the three-fault management methods under three-phase-to-ground fault.

Table 6.4 Performances of fault management methods under three-phase-to-ground fault.

	Distributed AFM	Naive ride-through method	Centralized AFM
Fault current contributions	19.41%	32.05%	20.03%

6.7   Conclusion

A novel distributed AFM is developed as a tool to ensure uninterrupted and reliable operations of networked microgrids during grids faults. Once a balanced or unbalanced fault occurs, the distributed AFM aims to (1) maintain the total fault current largely unchanged to avoid detrimental impact on the utility grid, (2) eliminate the damaging power ripples for inverters in the networked microgrids, and (3) ensure the power flow of each individual microgrid identical before and after fault to avoid loss of loads and maintain microgrid stability. The distributed AFM is modeled as a multiobjective optimization problem that can be solved online by an efficient DA-SLR method. The feasibility of the distributed AFM has been proved in a test case with six networked microgrids. In the future, the scalability of the distributed optimization solution is a major topic to be explored to enable an online, fast distributed AFM solution for very large-scale networked microgrids. Formal verification of AFM is also an important topic that needs to be further investigated to ensure the provably guaranteed performance of AFM under various uncertainties and disturbances.

References

[1] J. D. McDonald, "Microgrids beyond the Hype: Utilities Need to See a Benefit [Technology Leaders]," *IEEE Electrification Magazine*, vol. 2, no. 1, pp. 6–11, 2014.

[2] A. Hoke, S. Chakraborty, and T. Basso, "Testing Advanced Photovoltaic Inverters Conforming to IEEE Standard 1547-Amendment 1," in *2014 IEEE 40th Photovoltaic Specialist Conference (PVSC)*, IEEE, pp. 1014–1021, 2014.

[3] F. Zheng, C. Deng, L. Chen, S. Li, Y. Liu, and Y. Liao, "Transient Performance Improvement of Microgrid by a Resistive Superconducting Fault Current Limiter," *IEEE Transactions on Applied Superconductivity*, vol. 25, no. 3, pp. 1–5, 2015.

[4] Q. Wang, N. Zhou, and L. Ye, "Fault Analysis for Distribution Networks with Current-Controlled Three-Phase Inverter-Interfaced Distributed Generators," *IEEE Transactions on Power Delivery*, vol. 30, no. 3, pp. 1532–1542, 2015.

[5] A. Oudalov, T. Degner, F. v. Overbeeke, and J. M. Yarza, "Microgrid Protection," *Microgrids: Architectures and Control*, N. Hatziargyriou, ed., John Wiley and Sons, pp. 117–164, 2014.

[6] F. Nejabatkhah and Y. W. Li, "Overview of Power Management Strategies of Hybrid AC/DC Microgrid," *IEEE Transactions on Power Electronics*, vol. 30, no. 12, pp. 7072–7089, 2015.

[7] R. Salcedo, A. Bokhari, M. Diaz-Aguilo, N. Lin, T. Hong, F. De Leon, D. Czarkowski, S. Flank, A. McDonnell, and R. Uosef, Benefits of a Non-Synchronous Microgrid on Dense-Load Low-Voltage Secondary Networks," *IEEE Transactions on Power Delivery*, vol. 31, no. 3, pp. 1076–1084, 2016.

[8] M. A. Zamani, A. Yazdani, and T. S. Sidhu, "A Control Strategy for Enhanced Operation of Inverter-Based Microgrids under Transient Disturbances and Network Faults," *IEEE Transactions on Power Delivery*, vol. 27, no. 4, pp. 1737–1747, 2012.

[9] N. Rajaei, M. H. Ahmed, M. M. Salama, R. Varma, "Fault Current Management Using Inverter-Based Distributed Generators in Smart Grids," *IEEE Transactions on Smart Grid*, vol. 5, no. 5, pp. 2183–2193, 2014.

[10] S. I. Nanou and S. A. Papathanassiou, "Modeling of a PV System with Grid Code Compatibility," *Electric Power Systems Research*, vol. 116, pp. 301–310, 2014.

[11] E.-E. ENTSO-E, "Network Code for Requirements for Grid Connection Applicable to All Generators (NC RFG)," 2013. https://eepublicdownloads.entsoe.eu/clean-documents/pre2015/resources/RfG/130308_Final_Version_NC_RfG.pdf.

[12] Y. Wang, P. Zhang, W. Li, W. Xiao, and A. Abdollahi, "Online Overvoltage Prevention Control of Photovoltaic Generators in Microgrids," *IEEE Transactions on Smart Grid*, vol. 3, no. 4, pp. 2071–2078, 2012.

[13] W. Wan, Y. Li, B. Yan, et al., "Active Fault Management for Microgrids," *44th Annual Conference of the IEEE Industrial Electronics Society (IECON)*, pp. 1–6, 2018.

[14] W. Wan, Y. Li, B. Yan, M. A. Bragin, J. Philhower, P. Zhang, and P. B. Luh, "Active Fault Management for Networked Microgrids," in *2019 IEEE Power and Energy Society General Meeting (PESGM)*, Atlanta, Georgia, pp. 1–5, 2019.

[15] M. Di Somma, B. Yan, N. Bianco, G. Graditi, P. B. Luh, L. Mongibello, and V. Naso, "Operation Optimization of a Distributed Energy System Considering Energy Costs and Exergy Efficiency," *Energy Conversion and Management*, vol. 103, pp. 739–751, 2015.

[16] K. J. Devlin, *Fundamentals of Contemporary Set Theory*. Springer Science & Business Media, 2012.

[17] M. A. Bragin, P. B. Luh, and B. Yan, "Distributed and Asynchronous Coordination of a Mixed-Integer Linear System via Surrogate Lagrangian Relaxation," *IEEE Transactions on Automation Science and Engineering*, doi: 10.1109/TASE.2020.2998048.

[18] D. P. Bertsekas, *Nonlinear Programming*. Athena Scientific, 2016.

[19] X. Zhao, P. B. Luh, and J. Wang, "Surrogate Gradient Algorithm for Lagrangian Relaxation," *Journal of Optimization Theory and Applications*, vol. 100, no. 3, pp. 699–712, 1999.

[20] P. T. Boggs and J. W. Tolle, "Sequential Quadratic Programming," *Acta numerica*, vol. 4, pp. 1–51, 1995.

[21] Z. Zhang and M. Chow, "Convergence Analysis of the Incremental Cost Consensus Algorithm under Different Communication Network Topologies in a Smart Grid," *IEEE Transactions on Power Systems*, vol. 27, no. 4, pp. 1761–1768, 2012.

[22] M. A. Bragin and P. B. Luh, "Distributed and Asynchronous Unit Commitment and Economic Dispatch," *2017 IEEE Power Energy Society General Meeting*, pp. 1–5, 2017.

[23] MathWorks. Parallel Computing Toolbox. 2019 www.mathworks.com/products/parallel-computing.html.

7 Cyberattack-Resilient Networked Microgrids

7.1 Motivation

A communication network is essential in the aforementioned distributed calculation (DFA, DQG, ComPF); and thus, cybersecurity of a communication network is an indispensable concern. There are two essential constituents to be secured in order to guarantee cyber resilient NM operations. One is the communication infrastructure, and the other one is inverter controllers in DERs.

Here, for the communication infrastructure, we introduce software-defined networking (SDN) [1], a transformative technology for microgrid control and management pioneered through the author's recent work [2, 3]. SDN-based microgrid communication has gained popularity in recent years because it empowers ultrafast microgrid control, enables flexible networking of microgrids, and allows ultrafast response to NM contingencies. SDN provides logically centralized network intelligence and enables the microgrid communication infrastructure to be programmatically configured in real time, which has the potential to significantly improve the reliability, resiliency, and security of NMs [4]. However, due to its flexible configurability and a global access to the network [5, 6], it also makes the system potentially vulnerable to the so-called *first generation of cyberattacks* – meaning attackers aim to compromise the operations of power grids such as NMs by exploiting the information technology (IT) networks such as SDN. In addition, the newly emerging *second generation of cyberattacks*, such as *power bots*, poses significant threats on NMs and power systems in general. Power bot attack is a new type of cyberattack where remote attackers directly control or even corrupt power-consuming or DER devices to induce damages or blackouts in power systems. Therefore, cyberattacks are becoming one of the main hurdles to operate NMs.

Recently, several methods have been proposed in the computer science community to improve the awareness of cyberattacks against SDN. For example, a layer between control plane and data plane is introduced in VeriFlow to detect the network anomalies [7]. FlowChecker is created to locate intraswitch misconfiguration in one single encoded FlowTable [8] by using the formal verification techniques. Connection migrations and actuating triggers are introduced in AvantGuard to detect data-to-control-plane saturation attacks [9]. FRESCO [10] introduces OpenFlow-based modules for cyberattack detection and mitigation. ANCHOR improves SDN security by implementing global policies [11]. DELTA provides a cybersecurity assessment environment in which we can perform various SDN attacks for testing purposes [12].

In this chapter, we introduce a Software-Defined Active Synchronous Detection (SDASD) approach to defending against both the cyberattacks on an SDN network (a typical first generation of cyberattacks) and power bot attack on inverters (a typical second generation of cyberattacks). First, an effective defense strategy is devised and implemented in the SDN network controller to identify and mitigate cyberattacks on the SDN network of NMs. Second, with a secured SDN network, an active synchronous detection method is devised to accurately detect and localize power bot attacks on NMs by sending a probe signal to inverter controllers in NMs and checking the responses. There are several advantages of SDASD:

- SDASD offers a highly lightweight solution for real-time identification and mitigation of cyberattacks on both cyber and physical layers of NMs. It means that the security of SDN-enabled communication infrastructure or inverter controllers would not be compromised. Therefore, secure and ultrafast microgrid control schemes are ensured, which lays a solid foundation in securing stable operations of the low-inertia NMs.
- SDN controller is able to send programmable small signals to probe microgrid inverters, allowing real-time localization of inverter controllers compromised by power bots. Once power bot attacks are localized, remedial actions such as isolation and recovery can be taken immediately, which leads to high cyber resilience in NMs.
- The SDASD methodology has the potential to be generalized to secure various NM functionalities, such as active fault management and distributed power sharing, via the OpenFlow protocol. Therefore, it can serve as a useful security tool to safeguard the software or hardware functions that support programmable, self-configuring, self-protecting, and self-healing NMs.

7.2 Architecture of Software-Defined Active Synchronous Detection

Two layers are included in the generic SDASD framework. One is the network security layer, which is designed to protect SDN network components such as the data plane. The other is the infrastructure defense layer to protect physical NM components, such as the power bot defense for probing attacks on inverter controllers in DERs. Obviously, the infrastructure defense layer relies on a secured SDN network to deliver the probe signals used for probing DER controllers. Figure 7.1 shows the architecture of SDASD.

The functions of each block in Figure 7.1 are explained as follows:

- *Networked Microgrids Coordination Center (NMCC)*: The function of NMCC is to coordinate the operations of each microgrid in NMs. The probe signals used for the power bot defense layer are also generated at NMCC.

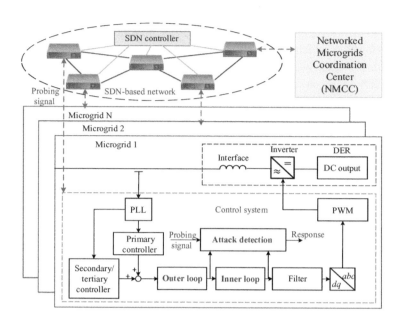

Figure 7.1 SDASD architecture.

- *SDN controller*: Protocols, such as OpenFlow [13], are used in SDN controller to enable network intelligence and centralized management of the communication infrastructure.
- *Microgrid*: When we have DC electricity from *DERs*, or a rectifier in a microgrid, an *inverter* will transform the DC electricity to AC. Each microgrid is then connected to a backbone network through an *interface* circuit to form the physical NMs . The function of the *control system* is to regulate the power output from DER. Inside the control system, the phase lock loop (*PLL*) is often used to identify the phase of input signals, which are usually the three-phase voltages. Then, in the *primary controller* or the *secondary/tertiary controller*, we use this phase to generate a droop-control signal and, if necessary, secondary or tertiary control signal, for the closed-loop controller, including *outer loop* and *inner loop*. A detection function is adopted in *attack detection* to identify attacks. Even though the impact of the probe signal to NM operation is negligible, a *filter* is still used to eliminate any potential noises induced. Since the reference values generated from the double-loop controller are *dq* signals, they need to be transformed into *abc* values, which can be taken by the *PWM* generator to create signals that operate power electronic switches, such as insulated-gate bipolar transistors (IGBTs), in the microgrid *inverter*.

A summary of the procedures for the detection and mitigation of attacks on the SDN network and inverter controllers is illustrated in Figure 7.2. More details for the cybersecurity techniques are detailed as follows.

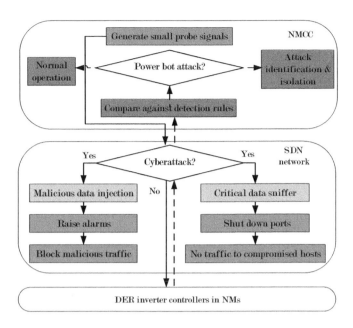

Figure 7.2 Defense steps against attacks on the SDN and power bot attacks on the inverter controller.

7.3   Defense against Cyberattacks on an SDN Network

In this chapter, our discussions focus on the defense strategies against the hijacking attacks of hosts that aim to exploit data communication between NMCC and microgrids. The reason is that attackers can relatively easily launch those attacks on IP-based NMs to poison data via hosts, thus they can immediately manipulate operations of an NM system and cause NM upset, collapse, information leakage, or various possible serious consequences. Even though those attacks on the SDN data plane seem daunting, there exists a straightforward and effective approach to defense against them: by leveraging the SDN controller functions, a *HostStatus_Checker* flag can be set up in the host tracking service (HTS) module. By doing so, the precondition and postcondition of hosts can be identified and thus the trustworthiness of the entity of these hosts can be indicated.

7.3.1   Update of the Host Tracking Service in an SDN Controller

In NMs, since the physical topology of microgrids may change frequently, the host or controller might migrate frequently as well. In the SDN controller, the host profile table is used to keep records of connected hosts and thus, authentication can be added to distinguish an intruder from a genuine host. Specifically, this can be achieved by monitoring the *HostStatus_Checker* flag inside the *Packet-In* entries via the HTS. If the HTS identifies the migration of a specific host to a different location, it updates the host profile table. Each entry of the host profile table consists of information such

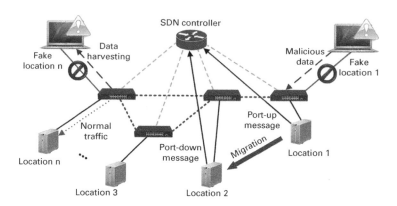

Figure 7.3 Cyberattack defending strategies.

as the host media access control (MAC) address, switch port number, and switch datapath ID (DPID) [5].

7.3.2   Defending Strategies

Illustrated in Figure 7.3 are the strategies to secure the SDN network, as follows:

- Pre event:

    Prior to the migration to a different location, a host of NMCC or microgrid will first send a *port-down* message to the SDN controller. Not only does this signal initiate the update process of the *HostStatus_Checker*, it also makes the original location of the host unreachable.

- Post event:

    Once the host successfully changes to a new location, it will send a *port-up* message to the SDN controller to inform the new location and thus complete the update of *HostStatus_Checker* and the host profile table. The host in the new location is now up for running.

- Defending actions against attacks:

    – Malicious data injection:

        An event will be identified as a host location hijacking attack if the received data packets have the same destination MAC/IP addresses but are from different locations that did not send *port-down* messages to the SDN controller beforehand. Once this event is identified, the system will first raise an alarm and then block the corresponding malicious traffic from those locations for the system's normal operation.

    – Critical data Snifferring:

        A host will be identified as a compromised host if the system detects that this host only sends *port-up* messages to the SDN controller but does not send any *port-down* messages. In this case, the corresponding port will be shut down immediately by the SDN controller. Further, the system will not send any data packets to the compromised host.

7.4   Active Synchronous Detection in DER Controllers of NMs

This section will present the use of active synchronous detection [14], which is extended and implemented in an SDN network, to detect the occurrence of power bot attacks in NMs. Basically, the following three steps can be established to enable the active synchronous detection:

- Step 1: Small probe signals are generated by the NMCC.
- Step 2: Through a secure SDN network, those signals are transferred to targets such as DER controllers.
- Step 3: When the targets receive those signals, they send back responses to the NMCC. The NMCC compares the received responses with predetermined detection rules, and identifies whether power bot attacks occur and where they occur.

7.4.1   Probe Signals for Active Synchronous Detection

Probe signals should be periodic and continuous and have small magnitudes in frequency domain, such that the targets' normal operations can be detected in real time and be nonimpediment. Those requirements can be mathematically described as follows:

$$s(t) = s(t + nT), \tag{7.1}$$

$$\|s(f)\| \leq \varepsilon, \tag{7.2}$$

$$\int_{t}^{t+T} s(t)dt = 0, \tag{7.3}$$

where $s(t)$ is a continuous signal in time domain, T is the time period of the signal, $\|s(f)\|$ is the l_2 norm of the signal's harmonic at frequency f, and ε is the threshold which is a small value. From (7.1) to (7.3), the following features can be obtained:

- A probe signal should have no impact on a target within one time period. In this way, the DER controllers' overall performance will not be changed by the probe signals.
- The NMCC should have the capability to program and flexibly modify the probe signals, if possible, such that the adversary's cost will be further increased.

7.4.2   Active Synchronous Detection on DER Controllers

As power bot attacks on DERs' inverter controllers can immediately compromise or collapse NMs, this type of power bot attack is considered the most malicious in NMs.

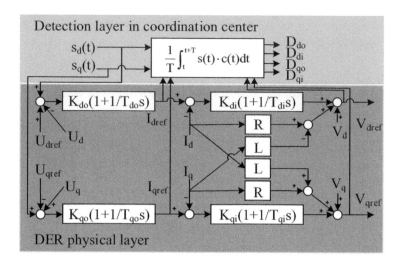

Figure 7.4 Illustration of the active synchronous detection on a double-loop controller.

To demonstrate how detection rules are built for power bot attacks on inverter controllers, the active synchronous detection of power bot attacks on a typical controller in DER inverters, the *dq* double-loop controller [15] (refer to Figure 7.1), is developed in this subsection.

The illustration of the active synchronous detection on a double-loop controller is given in Figure 7.4. $s_d(t)$ and $s_q(t)$ are probe signals generated in the NMCC. Through a secured SDN network, those signals are then delivered to critical DERs in each microgrid. Whenever necessary, those signals can be modified flexibly. The relationship between an attack detector signal and a probe signal can be represented as follows:

$$D = \frac{1}{T} \int_t^{t+T} s(t) \cdot c(t) dt, \tag{7.4}$$

where $s(t)$ represents a probe signal such as $s_d(t)$ and $s_q(t)$; $c(t)$ refers to a control signal such as I_{dref}, I_{qref}, V_{dref}, and V_{qref}; D is the corresponding attack detector signal, that is, an outer-loop signal such as D_{do} and D_{qo}, or an inner-loop signal such as D_{di} and D_{qi}. All the attack detector signals are computed within the NMCC, which will be discussed in the following subsection.

7.4.3   Detection Rules

To detect whether and where a power bot attack occurs, detection rules have to be derived and implemented in the NMCC. Typically, the following two types of power bot attacks can be considered:

(I) Controllers' topologies have been modified by the adversary.
(II) Controllers' parameters have been overwritten by the adversary.

To identify the type and location of a power bot attack on a certain DER controller, the first step is to generate probe signals in the NMCC and route them to that controller. The probe signals can be two sinusoidal signals, as shown in (7.5) and (7.6):

$$s_d(t) = \alpha_d \sin(\omega_d t), \tag{7.5}$$

$$s_q(t) = \alpha_q \sin(\omega_q t). \tag{7.6}$$

Within the NMCC, two detector signals D_{do} and D_{qo}, which are the outer-loop responses of the detection function (refer to (7.4)) in the dq coordinate, can be derived as shown in (7.7) and (7.8):

$$D_{do} = \frac{1}{T} \int_t^{t+T} s_d(t) I_{dref} dt = \frac{\alpha_d^2 K_{do}}{2}, \tag{7.7}$$

$$D_{qo} = \frac{1}{T} \int_t^{t+T} s_d(t) I_{qref} dt = \frac{\alpha_q^2 K_{qo}}{2}. \tag{7.8}$$

Similarly, two other detector signals D_{di} and D_{qi}, which are the inner-loop responses of the detection function (refer to (7.4)) in the dq coordinate, can be computed as shown in (7.9) and (7.10):

$$D_{di} = \frac{1}{T} \int_t^{t+T} s_d(t) V_{dref} dt = \frac{\alpha_d^2 K_{do} K_{di}}{2} \left(1 - \frac{1}{T_{do} T_{di} \omega_d^2} \right), \tag{7.9}$$

$$D_{qi} = \frac{1}{T} \int_t^{t+T} s_q(t) V_{qref} dt = \frac{\alpha_q^2 K_{qo} K_{qi}}{2} \left(1 - \frac{1}{T_{qo} T_{qi} \omega_q^2} \right). \tag{7.10}$$

For more details on the derivations of (7.7)–(7.10), readers are referred to [16]. In sum, the NMCC can detect power bot attacks on the outer loop of a DER controller via (7.7) and (7.8), and the attacks on the inner loop of a DER controller via (7.9) and (7.10). Specifically, the NMCC will determine whether a DER controller has been attacked or not by checking whether or not the received detector signals' steady-state values are identical to the computed values given by (7.7)–(7.10). The abnormal detector signals under the aforementioned two types of power bot attacks are summarized

Table 7.1 Values of detector signals in the d-axis under power bot attacks.

Controller under attack	Type of attacks	D_{do}	D_{di}
Outer loop	(1)	0	0
	(2)	$\alpha_d^2 K'_{do}/2$	$\alpha_d^2 K'_{do} K_{di}(T'_{do} T_{di}\omega_d^2 - 1)/2T'_{do} T_{di}\omega_d^2$
Inner loop	(1)	$\alpha_d^2 K_{do}/2$	0
	(2)	$\alpha_d^2 K_{do}/2$	$\alpha_d^2 K_{do} K'_{di}(T_{do} T'_{di}\omega_d^2 - 1)/2T_{do} T'_{di}\omega_d^2$
Outer loop	(1)	0	0
& inner loop	(2)	$\alpha_d^2 K'_{do}/2$	$\alpha_d^2 K'_{do} K'_{di}(T'_{do} T'_{di}\omega_d^2 - 1)/2T'_{do} T'_{di}\omega_d^2$

Table 7.2 Values of detector signals in the q-axis under power bot attacks.

Controller under attack	Type of attacks	D_{qo}	D_{qi}
Outer loop	(1)	0	0
	(2)	$\alpha_q^2 K'_{qo}/2$	$\alpha_q^2 K'_{qo} K_{qi}(T'_{qo}T_{qi}\omega_q^2 - 1 -)/2T'_{qo}T_{qi}\omega_q^2$
Inner loop	(1)	$\alpha_q^2 K_{qo}/2$	0
	(2)	$\alpha_q^2 K_{qo}/2$	$\alpha_q^2 K_{qo} K'_{qi}(T_{qo}T'_{qi}\omega_q^2 - 1 -)/2T_{qo}T'_{qi}\omega_q^2$
Outer loop & inner loop	(1)	0	0
	(2)	$\alpha_q^2 K'_{qo}/2$	$\alpha_q^2 K'_{qo} K'_{qi}(T'_{qo}T'_{qi}\omega_q^2 - 1 -)/2T'_{qo}T'_{qi}\omega_q^2$

in Tables 7.1 and 7.2. The location and type of a certain power bot attack can thus be determined by checking those abnormal values.

7.5 Test and Validation of Software-Defined Active Synchronous Detection

The NM system used to test and verify the performance of SDASD is given in Figure 7.5. Six microgrids are established and interconnected in this NM system, which operates in islanded mode with circuit breaker 0 open. The parameters of this test system can be found in Table 7.3.

The SDN topology designed in the test system is illustrated in Figure 7.6, where five SDN switches and one OpenFlow controller Ryu [17] are included. The NM model is developed in MATLAB/Simulink. The time step for the simulation is set at $50\,\mu s$, and the sample rate for the communication data is set at 10. A Mininet environment is utilized to run the SDN network. The bandwidth for each link is set at 1 Gbps in Mininet, which is commonly used in practice in an Ethernet network. To transfer information between the NMCC and microgrids through Mininet [18], the User Datagram Protocol (UDP) [19] can be used.

7.5.1 SDASD Performance Verification on Cyberattacks Defense

Initially, the NMCC runs in Center 1 with the IP address of 10.0.0.7. Center 2 is a newly migrated destination with the IP address of 10.0.0.8. Three test cases are conducted in this section to validate the effectiveness of SDASD on cyberattacks defense.

Case 1: Impact of Cyberattacks without SDASD
In this test case, Center 2 has been attacked in a way that it keeps sending out fake packets with the source IP and MAC addresses the same as those of Center 1, respectively. The fake packets are generated periodically through Scapy [20]. More specifically, before time $t = 2.0\,s$, there is no attack, meaning packets are only sent from

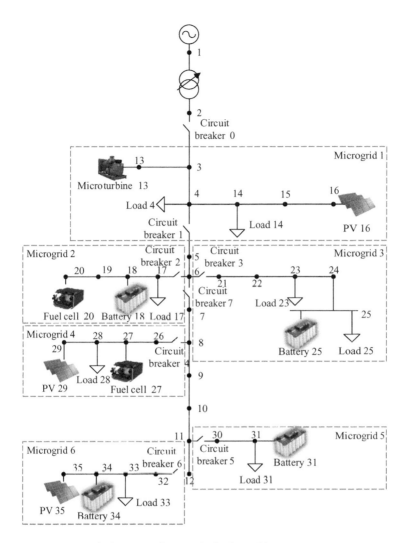

Figure 7.5 A typical system of networked microgrids.

Center 1. At time $t = 2\,s$, Center 2 is attacked and keeps injecting fake packets of probe signal $s_d(t)$ into Battery 25 with a payload of 10. Originally, the probe signal is a sinusoidal wave and has a frequency (ω_d) of 1,256 rad/s and an amplitude (α_d) of 0.01.

At time $t = 2.05\,s$, the abnormal value of D_{do} is detected, and immediately, Circuit Breaker 3 opens. The data traffic is shown in Figure 7.7. It can be seen that (1) before time $t = 2.0\,s$, all the packets received by Battery 25 are sent from Center 1; and (2) after time $t = 2.0\,s$, a certain number of packets are sent from Center 2 and the total number of packets increases. The probe signal $s_d(t)$ of Battery 25 before and after attacks is illustrated in Figures 7.8 and 7.9. It can be observed that the fake packets sent from Center 2 can maliciously impact normal traffics; more fake data are sent to

Table 7.3 Parameters of the test NM system in Figure 7.5.

Subsystems	From	To	$R(\Omega/km)$	$L(H/km)$	Length (m)	Bus	P_n (kW)	Q_n (kVAR)
Microgrid 1	3	4	0.2840	$0.2202e-3$	35	4	42.75	26.34
	3	13	0.2840	$0.2202e-3$	35	14	61.15	37.90
	4	14	3.6900	$0.2493e-3$	30	13*	98.15	64.02
	14	15	3.6900	$0.2493e-3$	30	16*	7.31	0.02
	15	16	3.6900	$0.2493e-3$	30			
Microgrid 2	17	18	1.3800	$0.2175e-3$	30	17	42.75	26.34
	18	19	1.3800	$0.2175e-3$	30	18*	17.45	25.76
	19	20	1.3800	$0.2175e-3$	30	20*	39.68	0.00
Microgrid 3	21	22	0.4970	$0.2281e-3$	30	25	61.15	37.90
	22	23	0.4970	$0.2281e-3$	30	23	61.15	37.90
	23	24	0.4970	$0.2281e-3$	30	25*	124.98	30.64
	24	25	0.8220	$0.2042e-3$	30			
Microgrid 4	26	27	0.8710	$0.2149e-3$	30	28	72.75	46.34
	27	28	0.8710	$0.2149e-3$	30	27*	39.70	0.00
	28	29	0.8710	$0.2149e-3$	30	29*	30.00	0.76
Microgrid 5	30	31	3.6900	$0.2493e-3$	30	31	62.75	57.91
						31*	69.40	30.50
Microgrid 6	32	33	1.3800	$0.2175e-3$	30	33	40.00	24.77
	33	34	1.3800	$0.2175e-3$	30	34*	1.65	148.48
	34	35	1.3800	$0.2175e-3$	30	35*	22.43	0.50
Backbone feeder	5	6	0.2840	$0.2202e-3$	35			
	7	8	0.2840	$0.2202e-3$	35			
	8	9	0.2840	$0.2202e-3$	35			
	9	10	0.2840	$0.2202e-3$	35			
	10	11	0.2840	$0.2202e-3$	35			
	11	12	0.2840	$0.2202e-3$	35			

*Denotes DER bus.

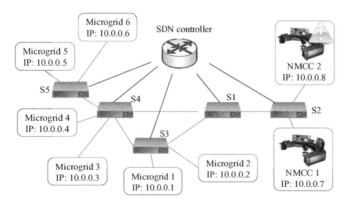

Figure 7.6 The topology of SDN network corresponding to the NM system in Figure 7.5.

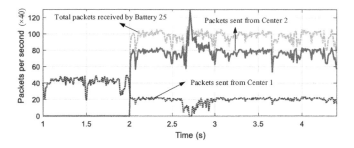

Figure 7.7 Case 1: Traffic monitoring of the packets received by Battery 25.

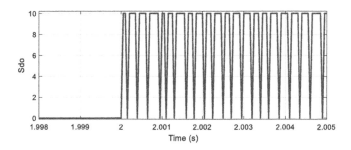

Figure 7.8 Case 1: An illustration of the probe signal $s_d(t)$ of Battery 25 before and after cyberattacks.

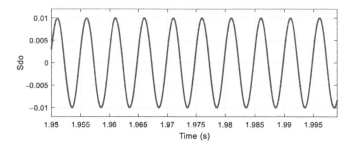

Figure 7.9 Case 1: An illustration of the probe signal $s_d(t)$ of Battery 25 before cyberattacks.

Battery 25. The voltage and current responses at bus 25 in Microgrid 3 are illustrated in Figure 7.10. It shows that the normal operations of the NM system can be severely impacted by cyberattacks, and the system can eventually collapse.

Case 2: Defending against Cyberattacks with SDASD

In this test case, the launched attack is the same as in Case 1. However, SDASD will be implemented to identify the anomaly in Center 2 and to protect the network for a secure data transmission. With SDASD, the SDN controller will receive traffic from malicious hosts through **packet-in** messages. However, no flow rules will be added, because there is no **port-down** message sent from those hosts to the SDN controller. Eventually, those malicious packets will be discarded directly. The raised alarm in the

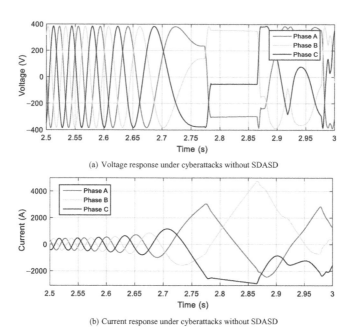

(a) Voltage response under cyberattacks without SDASD

(b) Current response under cyberattacks without SDASD

Figure 7.10 Case 1: Responses of the voltage and current at bus 25 in Microgrid 3.

```
loading app ryu/app/simple_switch_guard_13.py
loading app ryu.controller.ofp_handler
instantiating app ryu/app/simple_switch_guard_13.py of SimpleSwitch13
instantiating app ryu.controller.ofp_handler of OFPHandler
Need to check port down
Need to check port down
Need to check port down
```

Figure 7.11 Case 2: Alarm is raised in an SDN network.

SDN network is shown in Figure 7.11. Figure 7.12 gives an example of the normal operation of the system with SDASD.

The following can be observed from Figures 7.8 to 7.12:

- If there is no SDASD, a compromised host can easily inject malicious data packets into the SDN network, because from the SDN controller's point of view, the received packets that have the same source IP and MAC addresses are sent from the same location. However, as illustrated in Figure 7.7, a portion of the packets are actually sent from the compromised Center 2. Therefore, if there is no SDASD, the host location hijacking attack will pose a severe threat to the SDN network. Further, catastrophic collapse in the NM system will also be caused by such attacks as illustrated in Figure 7.10.

- If SDASD is implemented, malicious hijacking cyberattacks can be effectively blocked within the NM system. SDASD has the capability of achieving the system's reliable operations under certain cyberattacks, and possibly avoiding customers' economic losses.

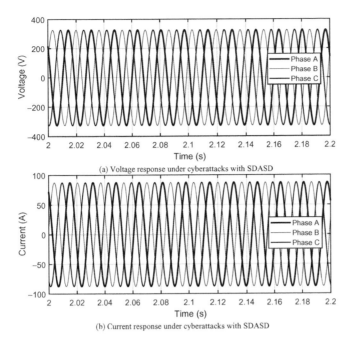

(a) Voltage response under cyberattacks with SDASD

(b) Current response under cyberattacks with SDASD

Figure 7.12 Case 2: Voltage and current responses at bus 25.

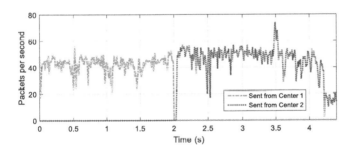

Figure 7.13 Case 3: Traffic monitoring during the process of normal migration with SDASD.

Case 3: Migration of a Normal NMCC with SDASD

The migration of a normal NMCC is demonstrated in this test case. At time $t = 2.0\,s$, the NMCC migrates from Center 1 to Center 2. Specifically, Center 1 sends a **port-down** message to the SDN controller, which preupdates its host profile table, resulting in the termination of sending packets from Center 1. Meanwhile, once Center 2 is initialized, it sends a **port-up** message to the SDN controller to update its host profile table. The SDN controller, when receiving those messages, will update flow rules in the switches, that is, add new rules and removing certain old rules. Eventually, the probe signals sent to Battery 25 will be from Center 2. In this case, normal operations will be maintained in the NM system as shown similarly to Figure 7.12. The traffic monitored in this process is illustrated in Figure 7.13.

The following can be observed from Figure 7.13:

- Before time $t = 2.0\,s$, all the probe signals received by Battery 25 are from Center 1; but after time $t = 2.0\,s$, they are all from Center 2. This indicates that NMCC has successfully migrated to a new center.
- A highly reliable host location migration can be guaranteed by SDASD. This is critical to enable resilient NM operations, as the system topology can be frequently changed due to common events in NM such as microgrid islanding and reconnecting.

7.5.2 Effectiveness of Active Synchronous Detection on Power Bot Attacks

SDASD is able to ensure a secure SDN network, where both probe signals and detection signals can be transferred securely. In this subsection, the following four cases are considered to validate the performance of active synchronous detection on power bot attacks, which are launched on DER controllers.

Case 4: Validation of SDASD on Type I Attack
At time $t = 1.10\,s$, the inner loop of the inverter on Battery 18 in Microgrid 2 is attacked by a type I attack. To validate the effectiveness of SDASD, two subcases, that is, running the test system with and without SDASD, are conducted in the test system. Specifically, when SDASD is established, α_d and α_q are both 0.01, and ω_d and ω_q are both 1,256 rad/s. The responses of the three-phase voltages at buses 18 and 25 without SDASD are illustrated in Figure 7.14, while those with

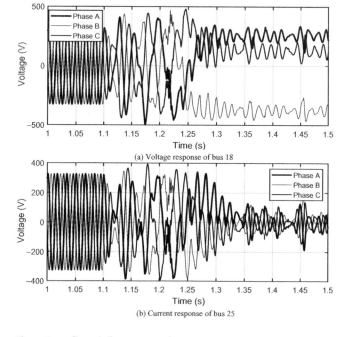

Figure 7.14 Case 4: Responses of the voltages at buses 18 and 25 with no SDASD.

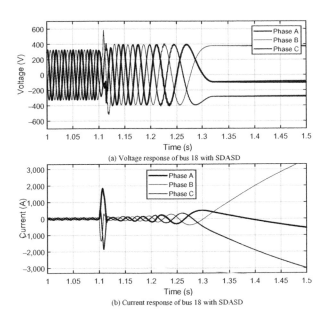

(a) Voltage response of bus 18 with SDASD

(b) Current response of bus 18 with SDASD

Figure 7.15 Case 4: Responses of voltage and current at bus 18 when SDASD is implemented.

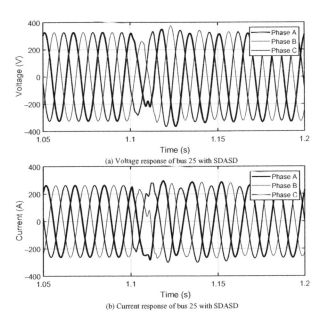

(a) Voltage response of bus 25 with SDASD

(b) Current response of bus 25 with SDASD

Figure 7.16 Case 4: Responses of voltage and current at bus 25 when SDASD is implemented.

SDASD are illustrated in Figures 7.15 and 7.16. The changes of the signal D_{do} in Battery 18 are given in Figure 7.17. The SDASD detects the power bot attack at time $t = 1.108$ when D_{do} is zero. Meanwhile, Circuit Breaker 2 opens immediately to isolate Microgrid 2.

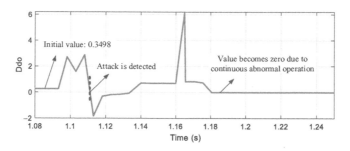

Figure 7.17 Case 4: Detected signal D_{do} for Battery 18.

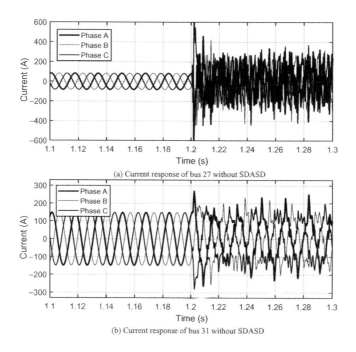

(a) Current response of bus 27 without SDASD

(b) Current response of bus 31 without SDASD

Figure 7.18 Case 5: Responses of the currents at buses 27 and 31 with no SDASD.

Case 5: Validation of SDASD on Type II Attack

At time $t = 1.20\,s$, a type II attack is launched on the inner loop of Fuel Cell 27's inverter in Microgrid 4. Specifically, the value of K_{di} is changed by the attacker from 0.25 to 10.0. The responses of the three-phase voltages at buses 27 and 31 without SDASD are illustrated in Figure 7.18, while those with SDASD are illustrated in Figures 7.19 and 7.19. At time $t = 1.217\,s$, the attack is detected by SDASD, and immediately, Circuit Breaker 4 opens to isolate Microgrid 4.

From Figures 7.14 to 7.20, the following can be observed:

- If there is no SDASD, the attack quickly affects the whole interconnected NMs and eventually collapses the system (refer to Figure 7.18).

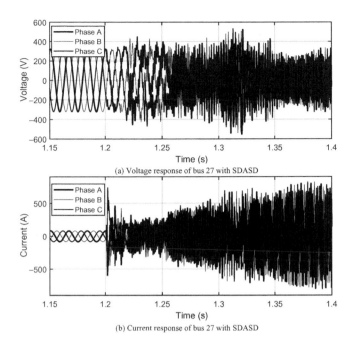

(a) Voltage response of bus 27 with SDASD

(b) Current response of bus 27 with SDASD

Figure 7.19 Case 5: Responses of voltage and current at bus 27 when SDASD is implemented.

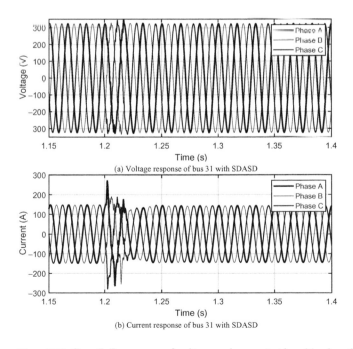

(a) Voltage response of bus 31 with SDASD

(b) Current response of bus 31 with SDASD

Figure 7.20 Case 5: Responses of voltage and current at bus 31 when SDASD is implemented.

- When SDASD is implemented, as shown in Figures 7.16 and 7.20, the system can detect and isolate the attack to mitigate the attack's impact. This verifies that SDASD has the capability of detecting power bot attacks.
- Before the system is attacked, the value of D_{do} in the experiment is 0.3498, as shown in Figure 7.17, which is approximately equal to the value calculated from (7.7). This validates that the detection rules are correct.
- D_{do} reaches zero (see Figure 7.17) when Microgrid 2 is isolated with the NM system. It seems to be inconsistent with the corresponding value given in Tables 7.1 and 7.2. This is because the attack causes an abnormal operation of Batter 18, while Tables 7.1 and 7.2 only provide steady-state detection results. Note that when detection results become abnormal, alarms need to be raised in practice.

Case 6: Multiattacks in Different Locations

In this test case, the inner loop of Battery 31's inverter is attacked by a type I attack at time $t = 2.10\,s$; at the same time, the inner loop of PV 35's inverter is attacked by a type II attack, where K_{di} is maliciously changed from 0.25 to 20.0. The responses of the three-phase voltages at bus 25 with and without SDASD are illustrated in Figure 7.21. It can be observed that with SDASD, the attack can be rapidly detected and isolated with the NM system. This verifies that SDASD is feasible to protect NMs from multiple attacks that occur simultaneously in different locations.

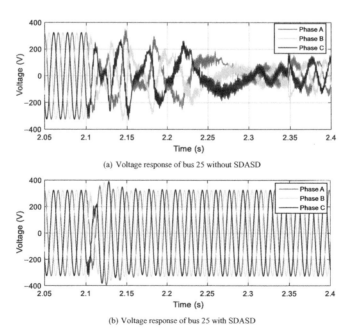

(a) Voltage response of bus 25 without SDASD

(b) Voltage response of bus 25 with SDASD

Figure 7.21 Case 6: Response of the voltage at bus 25 when multiple attacks are launched in different locations.

Case 7: Verification of SDASD's Robustness and Reliability

Robustness and reliability are two critical requirements that SDASD is expected to follow. It indicates that (1) SDASD should not malfunction during normal conditions when there is no attack; (2) for an arbitrarily switched probe signal, detection needs to be correct; and (3) SDASD should not affect normal NM operations. To validate SDASDs' robustness and reliability, both probe signals $s(t)$ and control signals such as U_{dref} in Figure 7.4 are changed online via NMCC. At time $t = 1.30\,s$, the magnitude reference signal of the voltage of Battery 34 in Microgrid 6 is modified from 0.99 p.u. to 0.90 p.u., that is, from 323.33 V to 293.94 V. At time $t = 1.42\,s$, the value is further changed to 0.95 p.u., that is, 310.27 V in the test system. For the probe signal $s_d(t)$, its amplitude is changed from 0.01 to 0.04 at time $t = 1.4\,s$, and is further changed to 0.02 at time $t = 1.50\,s$. The responses of the three-phase voltage and current at bus 34 with SDASD are illustrated in Figure 7.22, and the detection results of D_{do} in Battery 34 are illustrated in Figure 7.22.

The following can be seen from Figures 7.22 and 7.23:

- The changes of control signals can effectively dispatch DERs with no negative impact on detection rules as shown in the results between time $t = 1.30\,s$ and $t = 1.40\,s$. This verifies that SDASD is able to reliably distinguish between normal operations and operations under power bot attacks.

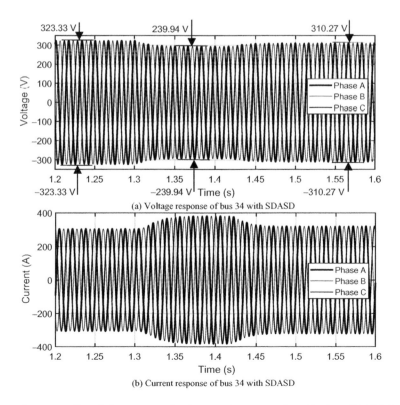

(a) Voltage response of bus 34 with SDASD

(b) Current response of bus 34 with SDASD

Figure 7.22 Case 7: Responses of voltage and current at bus 34 when SDASD is implemented.

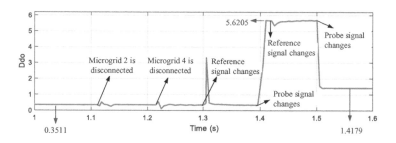

Figure 7.23 Case 7: Detected signal D_{do} for Battery 34.

- As shown in Figure 7.23, at time $t = 1.40\,s$, the detected signal changes from 0.3511 to 5.6205, that is, six times of 0.3511, while the amplitude of the probe signal quadruples from 0.01 to 0.04. At time $t = 1.50\,s$, the detected signal changes from 5.6202 to 1.4179, that is, 25% of 5.6205, while the amplitude of the probe signal reduces from 0.04 to 0.02, that is, half of 0.04. It verifies that the testing results are consistent with the detection rules shown in Tables 7.1 and 7.2.
- As illustrated in Figure 7.22, at time $t = 1.40\,s$ and $t = 1.50\,s$, normal operations are maintained in the test system during the adjusting process of the probe signal online, which indicates that SDASD is reliable with zero footprint on the test system.

Overall, SDASD neither requires adjusting the system's physical infrastructure, nor poses any perturbations on system states, and therefore provides a "non infrastructure" solution.

References

[1] D. Kreutz, F. M. Ramos, P. E. Verissimo, C. E. Rothenberg, S. Azodolmolky, and S. Uhlig, "Software-Defined Networking: A Comprehensive Survey," *Proceedings of the IEEE*, vol. 103, no. 1, pp. 14–76, 2015.

[2] L. Ren, Y. Qin, B. Wang, P. Zhang, P. B. Luh, and R. Jin, "Enabling Resilient Microgrid through Programmable Network," *IEEE Transactions on Smart Grid*, vol. 8, no. 6, pp. 2826–2836, 2017.

[3] P. Zhang, B. Wang, P. B. Luh, L. Ren, and Y. Qin, "Enabling Resilient Microgrid through Ultra-Fast Programmable Network," United States Patent. Patent No.: 10,505,853, Date of Patent: Dec. 10, 2019.

[4] M. Dabbagh, B. Hamdaoui, M. Guizani, and A. Rayes, "Software-Defined Networking Security: Pros and Cons," *IEEE Communications Magazine*, vol. 53, no. 6, pp. 73–79, 2015.

[5] S. Hong, L. Xu, H. Wang, and G. Gu, "Poisoning Network Visibility in Software-Defined Networks: New Attacks and Countermeasures." in *NDSS*, vol. 15, pp. 8–11, 2015.

[6] H. Farhady, H. Lee, and A. Nakao, "Software-Defined Networking: A Survey," *Computer Networks*, vol. 81, pp. 79–95, 2015.

[7] A. Khurshid, W. Zhou, M. Caesar, and P. Godfrey, "Veriflow: Verifying Network-Wide Invariants in Real Time," in *Proceedings of the First Workshop on Hot Topics in Software Defined Networks*. ACM, pp. 49–54, 2012.

[8] S. Al-Haj and W. J. Tolone, "Flowtable Pipeline Misconfigurations in Software Defined Networks," in *2017 IEEE Conference on Computer Communications Workshops (INFO-COM WKSHPS)*, IEEE, pp. 247–252, 2017.

[9] S. Shin, V. Yegneswaran, P. Porras, and G. Gu, "Avant-Guard: Scalable and Vigilant Switch Flow Management in Software-Defined Networks," in *Proceedings of the 2013 ACM SIGSAC Conference on Computer & Communications Security*. ACM, pp. 413–424, 2013.

[10] S. Shin, P. A. Porras, V. Yegneswaran, M. W. Fong, G. Gu, and M. Tyson, "FRESCO: Modular Composable Security Services for Software-Defined Networks," In *Proceedings of 2013 Network & Distributed System Security Symposium (NDSS'13)*, pp. 1–16, 2013.

[11] D. Kreutz, J. Yu, F. Ramos, and P. Esteves-Verissimo, "Anchor: Logically-Centralized Security for Software-Defined Networks," *arXiv preprint arXiv:1711.03636*, 2017.

[12] S. Lee, C. Yoon, C. Lee, S. Shin, V. Yegneswaran, and P. Porras, "DELTA: A Security Assessment Framework for Software-Defined Networks," in *Proceedings of 2017 Network & Distributed System Security Symposium NDSS'17)*, pp. 1–15, 2017.

[13] N. McKeown, T. Anderson, H. Balakrishnan, G. Parulkar, L. Peterson, J. Rexford, S. Shenker, and J. Turner, "Openflow: Enabling Innovation in Campus Networks," *ACM SIGCOMM Computer Communication Review*, vol. 38, no. 2, pp. 69–74, 2008.

[14] Y. Li, P. Zhang, L. Zhang, and B. Wang, "Active Synchronous Detection of Deception Attacks in Microgrid Control Systems," *IEEE Transactions on Smart Grid*, vol. 8, no. 1, pp. 373–375, 2017.

[15] C. Wang, Y. Li, K. Peng, B. Hong, Z. Wu, and C. Sun, "Coordinated Optimal Design of Inverter Controllers in a Micro-Grid with Multiple Distributed Generation Units," *IEEE Transactions on Power Systems*, vol. 28, no. 3, pp. 2679–2687, 2013.

[16] Y. Li, Y. Qin, P. Zhang, and A. Herzberg, "SDN-Enabled Cyber-Physical Security in Networked Microgrids," *IEEE Transactions on Sustainable Energy*, vol. 10, no. 3, pp. 1613–1622, 2019.

[17] S. Ryu, *Framework Community: Ryu SDN Framework*, https://book.ryu-sdn.org/en/Ryubook.pdf 2015.

[18] P. Chithaluru and R. Prakash, "Simulation on SDN and NFV Models through Mininet," in *Innovations in Software-Defined Networking and Network Functions Virtualization*. IGI Global, 2018, pp. 149–174.

[19] M.-H. Wang, P.-W. Chi, J.-W. Guo, and C.-L. Lei, "SDN Storage: A Stream-Based Storage System over Software-Defined Networks," in *2016 IEEE Conference on Computer Communications Workshops (INFOCOM WKSHPS)* IEEE, pp. 598–599, 2016.

[20] R. Montante, "Using Scapy in Teaching Network Header Formats: Programming Network Headers for Non-Programmers," in *Proceedings of the 49th ACM Technical Symposium on Computer Science Education*. ACM, pp. 1106–1106, 2018.

8 Networked DC Microgrids

8.1 Overview of DC Microgrids

Since the war of the currents in the late nineteenth century, AC distribution and transmission systems have been preferred over DC systems mainly due to the simple, high-efficiency AC transformers and polyphase AC machines. Although DC distribution and transmission systems were introduced as a simple solution for utilizing electric power at the beginning, they were not widely used due to the difficulties in DC voltage conversion and long-distance transmission. However, the development of the power electronic devices in the past decades offers efficient and low-cost solutions for DC voltage conversion and transmission, bringing DC power back to the stage, finding applications in high-voltage DC transmission (HVDC) lines, low-voltage distribution, data centers, and vehicle power systems [1, 2].

Recently, DC microgrids have been gaining popularity, mainly due to the rapid increase in the power-electronic loads and the proliferation of DC distributed energy resources (DERs) such as PV, fuel cell, and battery storage systems. In fact, as microgrids are replete with DC loads, it is economically and technically preferred to feed these loads directly by DC DERs, avoiding multiple conversion of voltage from DC to AC and vice versa [3, 4]. The following are other advantages of DC microgrids:

- Compared to AC microgrids, DC microgrids can offer a better power quality to the loads since they are not prone to reactive power imbalances and harmonic distortions.
- The control of DC microgrids is simpler as they need no frequency and reactive power regulations.
- The parallel operation of DC converters does not require synchronization, making DC microgrids more modular and flexible to expand.

Although the full implementation of DC microgrids is already happening in applications such as data centers and shipboard zonal power systems, the implementation of DC microgrids for low-voltage distribution systems has been impeded, mainly due to the fact that the current infrastructure has been built based on AC systems. Besides, for low-voltage application, there is no inexpensive, efficient solution for AC/DC power conversion as compared to simple yet efficient AC distribution transformers. However, there is ongoing research on the solid-state transformers to make it feasible for distribution systems.

In addition, the development of an appropriate protection scheme is still a major challenge in DC microgrids. The lack of natural zero-crossing current makes the interruption of fault current more difficult. Hence, the designing of DC circuit breakers is more challenging. The methods implemented for DC protection can be broadly classified as (1) the conventional circuit breakers equipped with resonance circuitry to generate zero-crossing DC fault current, and (2) power electronics based protective devices. The former uses inexpensive mechanical circuit breakers, but their fault current isolation speed is slow, typically in the range of 30–100 ms. The latter, however, is very fast in detection and isolation of a faulty circuit, but it has high power losses in the steady-state condition [1].

8.2 Bipolar DC Microgrids

Compared with the conventional (unipolar) DC microgrid, a bipolar-type DC microgrid provides additional advantages. Thanks to its three-wire topology, two voltage levels are available to the loads, meaning that heavy and light loads can be connected to higher and lower voltage levels, which not only reduces the ohmic losses, but also simplifies power converter topologies [5, 6].

The voltage level of a microgrid greatly depends on the diversity and power ratings of the electric loads; heavy loads, such as drive systems, require higher voltage levels to reduce ohmic losses and voltage drop, whereas light loads need lower voltage levels to simplify converter topology. Hence, microgrids should provide two or three voltage levels to the loads.

Voltage conversion can be easily achieved via inexpensive but efficient power transformers in AC microgrids, while this is a great challenge in DC microgrids. A typical structure of a conventional DC microgrid with the rated DC voltage of 800 V is shown in Figure 8.1a. High power loads are connected to 800 V DC busbars while low power loads are connected to the 400 V DC busbars. The interlink DC/DC converter responsible for reducing voltage from 800 to 400 V DC must be rated to supply power to the connected loads, which causes extra losses and cost. This can be avoided using bipolar DC microgrid. In fact, one of the advantages of the bipolar DC microgrid lies in its ability to provide two voltage levels without an interlink DC/DC converter, as it is shown in Figure 8.1b. In addition, only one storage system is required.

However, similar to AC microgrid, the bipolar DC microgrid is prone to load unbalance. Hence, a voltage balancer is required to balance the main busbar voltages under any loading conditions. Several DC/DC converter topologies have been introduced in the current literature to perform voltage-balancing function [7, 8].

The structure of a bipolar DC microgrid is similar to three-wire single phase system, which is common in North America, with the voltage levels of 120 V AC line-to-neutral (grounded center tap of transformer) and 240 V AC line-to-line. Using the same terminology, line-to-line voltage of 800 V DC would result in a 400 V DC line-to-neutral in a bipolar DC microgrid. The three-wire single phase AC system and a bipolar DC microgrid are compared in Figure 8.2.

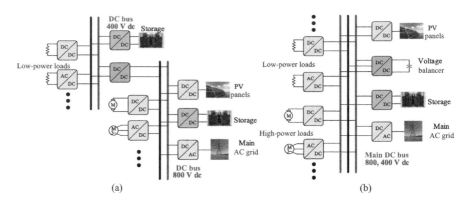

Figure 8.1 Typical structures of (a) a conventional, unipolar DC microgrid, and (b) a bipolar DC microgrid.

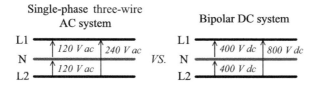

Figure 8.2 The similarity between a single-phase three-wire AC network and a bipolar DC microgrid.

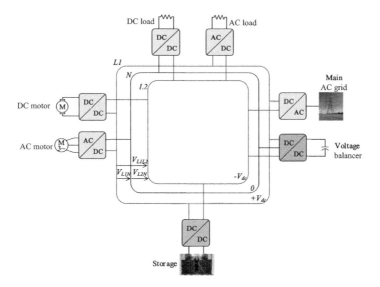

Figure 8.3 Ring-type bipolar DC microgrid.

The reliability and efficiency of bipolar DC microgrid can be further increased by using ring-type topology, as indicated in Figure 8.3. This structure is suitable for shipboard application, where the reliability and efficiency are of paramount importance.

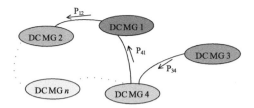

Figure 8.4 Networked DC microgrid.

8.3   Networked DC Microgrids

In case of islanded mode of operation or connection to a weak AC grid, DC micro-grids are unable to support their loads merely from renewable resources. Hence, a sudden load or generation change may easily lead to DC voltage instability. This can be avoided by interconnection of nearby individual DC microgrids in order to form networked DC microgrids [9].

As shown in Figure 8.4, the networking of individual DC microgrids can be realized by electrically connecting the DC or AC bus of individual DC microgrids. The networking of DC microgrids enables the maximum utilization of energy resources as it allows shared generation assets among loads, improves overall reliability, and can reduce maintenance costs. In terms of dynamics of the networked DC microgrids, the overall stability of the system may improve; however, a detailed analysis must be conducted to ensure local and system-level stability. In particular, the individual DC microgrids may be locally stable, but after networking, the overall stability might be reduced due to the mutual interactions between the control systems of DC microgrids.

8.4   Dynamic Modeling of DC Microgrids

DC microgrid is a converter-based power network that offers several benefits, such as excellent load regulation, improved transient performance, and fault tolerance. However, it is prone to voltage instability due to the interactions among the control loops of power converters.

In general, the DC voltage of the microgrid is controlled by the voltage control loops of source-side converters. This voltage, however, acts as disturbance to the control loops of load-side converters. In addition, the currents of the load-side converters act as disturbances to the control loops of source-side converters. The mutual interactions between the control loops of the source-side (AC/DC) and load-side (DC/DC) converters are illustrated in Figure 8.5. As can be seen, the control loops of the load-side converters also interact with each other. For instance, a change in the load current i_{load1} would also change the converter input current, i_{in1}. Subsequently, a change in i_{in1} results in the variations in v_{DC} as i_{in1} is a disturbance to the control system of the source-side converter that controls v_{DC}. Finally, the variations in v_{DC} would change

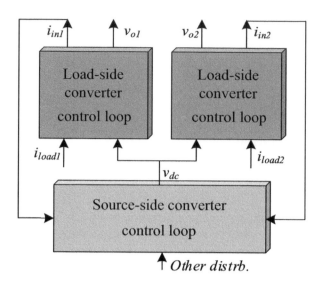

Figure 8.5 The interactions among the control loops of converters.

the output voltage v_{02}. Thus, if the interactions among voltages and currents of all converters are properly identified, the small-signal stability of the microgrid can be understood.

To address this challenge, each component in the DC microgrid is modeled via a Multi-Input Multi-Output (MIMO) transfer function matrix. Then the MIMO models are connected together based on the similar input/output signals. From the result MIMO model, which represents the overall dynamics of the DC microgrid, the small-signal stability and the mutual interactions between desired input–output pairs can be detailed.

In a bipolar DC microgrid, the control loops of converters interact through L1-N and L2-N voltages. In fact, v_{L1L2} is regulated via interlinking converter. Then, a voltage balancer controls v_{L1N} and v_{L2N}. These voltages are disturbances to the control loops of the load-side converters. Furthermore, any load change on the L1-N terminal would have impact on the L2-N terminal voltage and vice versa. This is due to the mutual coupling between the L1N and L2N terminals, which complicates the stability analysis. The line impedances are ignored to simplify the analysis, focusing on the interaction analysis; however, they can be also considered in the analysis.

The small-signal model of a bipolar DC microgrid is shown in Figure 8.6a. The symbol of $\triangle$ is used for signals to emphasize that the system is presented in small-signal variations. Every block in this diagram represents a component in the bipolar DC microgrid. A conventional two-level inverter is used as an interlinking converter regulating the DC line-to-line (L1–L2) voltage. The loads with high-rated power, such as AC or DC electric motor derives, are connected to v_{L1L2}. This voltage is then split to v_{L1N} and v_{L2N} to supply low power loads.

Every block in the diagram presented in Figure 8.6a is a MIMO model defined by a transfer function matrix. In the source side, the generating unit consists of an AC grid and AC/DC converter. The v_{L1L2} does not directly impact the control loops of

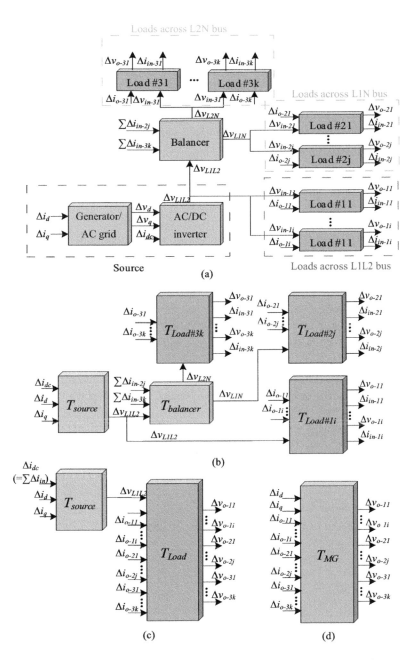

Figure 8.6 Bipolar DC microgrid small-signal model; various stages to obtain one overall MIMO model for the entire bipolar DC microgrid: (a) small-signal model of microgrid where each component is presented by a MIMO model; (b) the individual MIMO models are locally combined to build blocks of loads and a block of source; (c) the MIMO models are further combined to form load and source subsystems; and (d) overall MIMO model of the microgrid.

the loads connected to L1N and L2N terminals. It acts as a disturbance to the control system of the voltage balancer and the loads connected to L1L2 terminal. All loads in the bipolar DC microgrid, regardless of their types, should be modeled via the transfer function matrix in which the input voltage and output current (load current) are considered as the input variables; and the output voltage (load voltage) and input current are selected to be the output variables.

The voltage balancer is a key converter, and proper function of a bipolar DC microgrid is impossible without it. As shown in Figure 8.6a, the interaction among L1N and L2N loads occur through this converter. It is important to notice that the sum of the input currents of the loads connected across L1N bus, as well as the loads connected across L2N bus, act as disturbance to the voltage control loop of the balancer.

The loads connected to the bipolar DC microgrid can be classified as L1L2 loads (Load #11 to Load #1j), L1N loads (Load #21 to Load #2j), and finally L2N loads (Load #31 to Load #3j). The subsystems illustrated in Figure 8.6a can be combined to achieve the system shown in Figure 8.6b. The transfer matrix T_{source} can reproduce the dynamics of the complete generating sets, including the AC/DC inverter.

The system presented in Figure 8.6b can be regarded as two subsystems: source (T_{source}) and load (T_{Load}), which is shown in Figure 8.6c. Although $T_{balancer}$ is neither load nor source, it is integrated into the load subsystem. The final step is to combine the two subsystems to obtain the overall MIMO model of the bipolar DC microgrid, which is shown in Figure 8.6d. The transfer function matrix T_{MG} reproduces the overall dynamics of the bipolar DC microgrid.

8.4.1   Implementation

Without loss of generality, the MIMO modeling approach is applied to the bipolar DC microgrid presented in Figure 8.7. The given bipolar DC microgrid consists of three loads, a two-level inverter, and a voltage balancer. Load #1, Load #2, and Load #3 are connected to L1L2 bus, L1N bus, and L2N bus, respectively. The transfer function matrices of the loads have the following equation,

$$\begin{bmatrix} \triangle v_o \\ \triangle i_{in} \end{bmatrix} = T_{Load} \begin{bmatrix} \triangle v_{in} \\ \triangle i_o \end{bmatrix}, \tag{8.1}$$

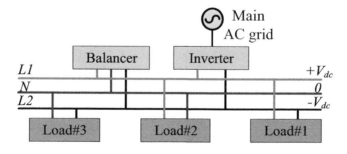

Figure 8.7 Bipolar DC microgrid for MIMO stability analysis.

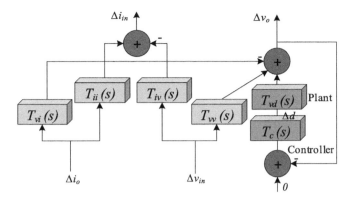

Figure 8.8 Control system of conventional DC/DC converters.

where T_{Load} is a 2×2 transfer function matrix. The load can be a conventional DC/DC converter or DC/AC electric motor drive. Regardless of the load types, loads can be expressed by the equation given by (8.1).

The control system of a DC/DC converter is shown in Figure 8.8. The output current (load current), Δi_o, and the input voltage, Δv_{in}, are considered as disturbances to the output DC voltage. There are five plant transfer functions to be calculated, including $T_{vd}(s)$, $T_{vv}(s)$, $T_{iv}(s)$, $T_{ii}(s)$, and $T_{vi}(s)$, and one PI controller, $T_c(s)$. Using the state-space averaging method, the general small-signal presentation for the common DC/DC converters can be expressed as

$$\dot{\Delta x} = A_{ave} \Delta x + B_{ave} \Delta u + [(A_1 - A_2)X + (B_1 - B_2)U]\Delta d \tag{8.2}$$

$$\Delta y = C_{ave} \Delta x + E_{ave} \Delta u + [(C_1 - C_2)X + (E_1 - E_2)U]\Delta d, \tag{8.3}$$

where x (state vector) contains capacitor voltages and inductor currents and u (input vector) contains input DC voltage and output current. The matrices X and U contain the steady-state values of state variables and inputs, respectively. A_1 to E_1 and A_2 to E_2 are coefficient matrices. The averaged matrices are indicated with subscript ave. The converter duty cycle is d, and the desirable output variables are in the vector y. Adopting (8.2) and (8.3), the transfer functions of a conventional DC/DC converter can be derived by

$$\Delta x = (s\,I - A_{ave})^{-1} ((A_1 - A_2)X + (B_1 - B_2)U)\Delta d \tag{8.4}$$

$$\Delta x = (s\,I - A_{ave})^{-1} B_{ave}\Delta u. \tag{8.5}$$

Taking a DC/DC boost converter as an example, the capacitor voltage and inductor current would be regarded as, respectively, output voltage and input current. Hence,

$$\Delta x = [\Delta v_o, \Delta i_{in}]' \ and \ \Delta u = [\Delta v_{in}, \Delta i_o]'$$

$$A_1 = \begin{bmatrix} 0 & -1/CR \\ 0 & 0 \end{bmatrix}, \ B_1 = \begin{bmatrix} 1/L & 0 \\ 0 & -1/C \end{bmatrix}$$

$$A_2 = \begin{bmatrix} 1/C & -1/CR \\ 0 & -1/L \end{bmatrix}, \ B_2 = \begin{bmatrix} 1/L & 0 \\ 0 & -1/C \end{bmatrix}$$

$$A_{ave} = d\,A_1 + (1-d)A_2, \ B_{ave} = d\,B_1 + (1-d)B_2,$$

where C, L, and R are, respectively, output capacitor, inductor, and load resistance of the boost converter. The plant transfer function $T_{vd}(s)$ can be calculated by (8.4), and the rest of transfer functions are obtained from (8.5) as follows:

$$T_{vd} = [1,0] \times [(s\,I - A_{ave})^{-1}((A_1 - A_2)X + (B_1 - B_2)U)]$$

$$\begin{bmatrix} T_{vv} & -T_{vi} \\ -T_{iv} & T_{ii} \end{bmatrix} = [(s\,I - A_{ave})^{-1}\,B_{ave}].$$

After computing the transfer functions of Figure 8.8, the transfer function matrix T_{Load} can be built:

$$T_{Load} = \begin{bmatrix} \dfrac{T_{vv}}{1+T_c*T_{vd}} & \dfrac{-T_{vi}}{1+T_c*T_{vd}} \\ -T_{iv} & T_{ii} \end{bmatrix}. \tag{8.6}$$

The first row of T_{Load}, which is related to the output voltage, implies a closed-loop form. This reflects that the converter is operating in DC voltage control mode.

The transfer function matrix of the voltage balancer, $T_{balancer}$, can be obtained using the same modeling approach just described. Here again, the disturbances to the control system of the voltage balancer are load currents (i_{oL1} and i_{oL2}) and the input voltage (v_{L1L2}). The control system of the balancer maintains the difference between the output voltages, v_{L1N} and v_{L2N}, at zero. v_{L1L2} is controlled via inverter. The dynamics of the voltage balancer and its control system can be detailed using eight plant transfer functions and one PI controller (see Figure 8.9). The output capacitor

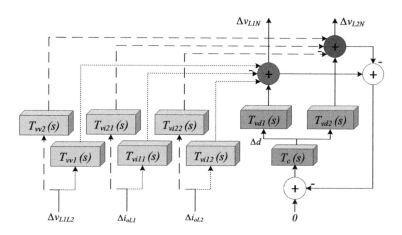

Figure 8.9 Control system of voltage balancer.

voltages and inductor current are the elements of $\triangle x$. The capacitor voltages are equal to L1N and L2N voltages. So,

$$\triangle x = [\triangle v_{L1N}, \triangle v_{L2N}, \triangle i_L]'$$

$$\triangle u = [\triangle i_{oL1}, \triangle i_{oL2}, \triangle v_{L1L2}]'$$

$$A_1 = \begin{bmatrix} -1/2Cr_c & -1/2Cr_c & 1/2C \\ -1/2Cr_c & -1/2Cr_c & -1/2C \\ -1/2L & 1/2L & (r_c)/2L \end{bmatrix}$$

$$B_1 = \begin{bmatrix} -1/2C & 1/2C & 1/2Cr_c \\ -1/2C & -1/2C & 1/2Cr_c \\ r_c/2L & -r_c/2L & 1/2L \end{bmatrix},$$

where L, C, and r_c are inductor, output capacitor, and the associated equivalent series resistance, respectively (see [10] for the balancer topology). A_2 and B_2 are equal to A_1 and B_1, respectively, except that $B_1(3,3) = -B_2(3,3)$. The plant transfer functions can be calculated from (8.4) and (8.5):

$$\begin{bmatrix} T_{vd1} \\ T_{vd2} \end{bmatrix} = \begin{bmatrix} 1,0,0 \\ 0,1,0 \end{bmatrix} [(sI - A_{ave})^{-1}((A_1 - A_2)X + (B_1 - B_2)U)]$$

$$\begin{bmatrix} -T_{vi11} & T_{vi12} & T_{vv1} \\ T_{vi21} & -T_{vi22} & T_{vv2} \\ 0 & 0 & 0 \end{bmatrix} = \begin{bmatrix} 1,1,1 \\ 1,1,1 \\ 0,0,0 \end{bmatrix} [(s\,I - A_{ave})^{-1} B_{ave}].$$

The balancer is described by the following equation set:

$$\begin{bmatrix} \triangle v_{L1N} \\ \triangle v_{L2N} \end{bmatrix} = T_{balancer} \begin{bmatrix} \triangle i_{oL1} \\ \triangle i_{oL2} \\ \triangle v_{L1L2} \end{bmatrix}. \tag{8.7}$$

The small-signal model of a two-level AC/DC inverter that connects the AC grid to the bipolar DC microgrid is presented in Figure 8.10. The active power equation can be linearized as

$$\triangle p_s = \frac{3}{2} I_d \,\triangle v_d + \frac{3}{2} V_d \,\triangle i_d, \tag{8.8}$$

where I_d and V_d are steady-state current and voltage, respectively. The inverter operates in DC voltage control mode, meaning v_{L1L2} is controlled by a DC voltage control loop, which in fact regulates the stored energy of the DC-link capacitor. Since electrical energy is proportional to the square of voltage across DC-link capacitor, it is simpler to control the DC voltage using its square value. The voltage control loop provides the active power reference of the inverter, p_s^{ref}. Then, it is used to generate the inner current control loop in dq frame.

In the grid-connected mode, the inverter exchanges power with the AC grid while maintaining v_{L1L2} at the V_{L1L2}^{ref} value. The AC-side voltage and frequency of the

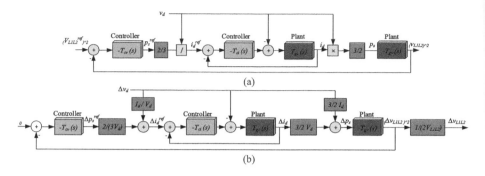

Figure 8.10 Control system of the AC/DC inverter in DC-link voltage control mode, (a) nonlinear model, and (b) linear model.

inverter are imposed by the AC grid, while the inverter injects the active power of p_s^{ref} into the AC grid.

The transfer function relating $\triangle v_{L1L2}$ to the d-axis component of AC grid voltage ($\triangle v_d$) is obtained from the linear model presented in Figure 8.10b. Since the main AC grid is stronger as compared with the bipolar DC microgrid, the dynamics of the AC side can be simplified. Therefore, the source transfer function will have the following equation:

$$T_{source}(s) = \frac{\triangle v_{L1L2}(s)}{\triangle v_d(s)}. \tag{8.9}$$

Referring to Figure 8.6, the transfer function matrices given by (8.1), (8.7), and (8.9) can be integrated to study the small-signal stability of the bipolar DC microgrid. Note that $\triangle v_{L1N}$ and $\triangle v_{L2N}$, maintained by the voltage balancer, are disturbances to the control systems of the load converters. Moreover, the disturbances to the control system of the voltage balancer are $\triangle i_{oL1}$ and $\triangle i_{oL2}$, which are the sum of the load currents connected to, respectively, the L1N bus and L2N bus. Thus, the transfer function matrix of the voltage balancer, which also includes the dynamics of the loads connected across the L1N and L2N bus, is expressed as

$$\begin{bmatrix} \triangle v_{L1N} \\ \triangle v_{L2N} \end{bmatrix} = T_{bal} \begin{bmatrix} \triangle i_{o2} \\ \triangle i_{o3} \\ \triangle v_{L1L2} \end{bmatrix}. \tag{8.10}$$

The coupling effects (interaction) between the L1N and L2N bus can be observed by (8.10). In fact, adopting (8.10), the impact of a load change on the L2N bus can be identified on the output voltage of the load connected to the L1N bus. Next, the transfer function matrix of the voltage balancer is integrated into the load transfer function matrix,

$$\begin{bmatrix} \triangle v_{o1} \\ \triangle v_{o2} \\ \triangle v_{o3} \end{bmatrix} = T_{Load} \begin{bmatrix} \triangle i_{o1} \\ \triangle i_{o2} \\ \triangle i_{o3} \\ \triangle v_{L1L2} \end{bmatrix}. \tag{8.11}$$

Finally, the dynamics of the inverter are inserted into (8.11) in order to establish the overall MIMO model that reproduces the dynamics of the bipolar DC microgrid,

$$
\begin{bmatrix} \Delta v_{o1} \\ \Delta v_{o2} \\ \Delta v_{o3} \end{bmatrix} = T_{MG} \begin{bmatrix} \Delta i_{o1} \\ \Delta i_{o2} \\ \Delta i_{o3} \\ \Delta v_d \end{bmatrix},
\tag{8.12}
$$

where $T_{MG}(s)$ is a transfer function matrix with an order of 3 by 4, which provides a suitable mean for stability and mutual interaction analysis.

8.4.2   MIMO Tools for Stability and Interaction Analysis

Once $T_{MG}(s)$ is obtained, the stability of the bipolar DC microgrid and the mutual interactions among various input–output pairs can be studied. The Geršgorin circle (Geršgorin band) is used to reveal the mutual interactions and the degree of coupling. The Geršgorin band specifies a region in the complex plane that contains all eigenvalues of a given complex matrix. For an operating point, $T_{MG}(jw)$ is a complex matrix whose eigenvalues fall on the union of circles with the center at $T_{MG}(i, i)$ and radius

$$
\sum_{j=1, j \neq i}^{4} |T_{MG}(i, j)|
\tag{8.13}
$$

and also on the circles with the center at $T_{MG}(j, j)$ and radius

$$
\sum_{i=1, i \neq j}^{3} |T_{MG}(i, j)|.
\tag{8.14}
$$

The aforementioned circles are superimposed on the Nyquist diagrams of $T_{MG}(i, i)$. Note that the result Nyquist diagrams are not used for stability analysis; instead, the regions occupied by Geršgorin bands are of interest for the interaction analysis. If the Geršgorin bands are thin, the $T_{MG}(s)$ is diagonally dominant, which is interpreted as a weakly coupled MIMO system. In other words, the Geršgorin theorem reveals how close a MIMO transfer function matrix is to being diagonal. The more diagonal it is, the weaker interactions between L1N and L2N.

The frequency-dependent Relative Gain Array (RGA) analysis is also utilized as a complement to the Geršgorin theorem. RGA analysis quantifies the mutual interactions among control loops in the frequency range of interest. The stronger interactions (coupling) among control loops of converters are associated with the higher RGA values. In control engineering, RGA is defined as

$$
RGA(T_{MG}(jw)) \triangleq T_{MG}(jw) \times ((T_{MG}(jw))^{-1})^{T},
\tag{8.15}
$$

where the transfer function matrix $T_{MG}(s)$ is multiplied by its inverse transposed matrix at the frequency of interest.

The Geršgorin theorem identifies the degree of coupling among desired input–output pairs; however, it is unable to give information about how large the system response is for specific inputs. In MIMO systems, singular value decomposition (SVD) provides a system frequency response that shows how large the output amplification is for a specific vector of inputs.

The singular values of the microgrid transfer function matrix, $T_{MG}(s)$, for $s = jw$, are denoted as

$$\sigma_i \left(T_{MG}(jw) \right) = \sqrt{\lambda_i \left(T_{MG}(jw)^T \, T_{MG}(jw) \right)}, \qquad (8.16)$$

where $\lambda_i(\cdot)$ denotes the ith eigenvalue of the matrix. The maximum amplification that the output vector can experience is given by the maximum singular value $\overline{\sigma}(T_{MG}(jw))$ and the minimum amplification by the minimum singular value $\underline{\sigma}(T_{MG}(jw))$.

8.5   Stability and Mutual Interactions Analysis

The bipolar DC microgrid presented in Figure 8.7 has a rated line to neutral voltage of 400 Vdc and line-to-line voltage of 800 Vdc. The Load#1 is a DC/DC boost converter loaded by a resistive load; while Load#2 and Load#3 are DC/DC buck converters, which are again loaded by resistive loads. The microgrid specifications and parameters are given by Table 8.1, and the control systems implemented by the converters are shown in Figures 8.8–8.10.

8.5.1   Local Interactions

This study details how interactions can cause a locally stable control system to become unstable when connected to the bipolar DC microgrid.

In this study, Load#1, Load#2, and Load#3 consume power of, respectively, 10 kW, 2 kW, and 2 kW, meaning the microgrid is balanced. In steady-state conditions, the

Table 8.1 Bipolar DC microgrid parameters.

Item	Specifications
Load#1	$V_{out} = 1{,}000$ V, $P_{nom} = 10$ kW, $L = 3.7$ mH, $C = 10$ μF, $f_s = 40$ kHz
Load#2	$V_{out} = 100$ V, $P_{nom} = 2$ kW, $L = 938$ μH, $C = 313$ μF, $f_s = 40$ kHz
Load#3	$V_{out} = 200$ V, $P_{nom} = 8$ kW, $L = 625$ μH, $C = 625$ μF, $f_s = 40$ kHz
Inverter	$P_{nom} = 20$ kW, $V_{L1L2} = 800$ V, $L = 2$ mH, $r_L = 2$ mΩ, $C = 20$ mF, $f_s = 4$ kHz
Balancer	$V_{out} = 2 \times 400$ V, $P_{nom} = 3$ kW, $L = 600$ μH, $C = 5$ mF, $f_s = 40$ kHz
AC grid	$v_{L-L} = 380$ $Vrms$, $f = 60$ Hz

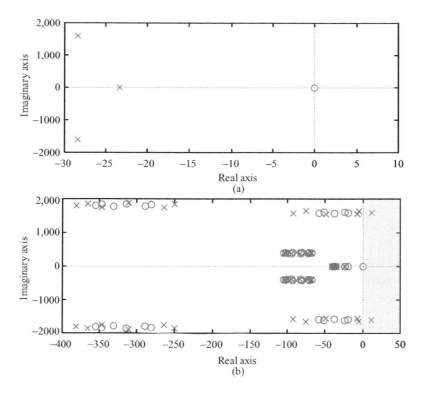

Figure 8.11 Pole-zero map of $\triangle v_{o3}/\triangle i_{o3}$: (a) before connecting to the microgrid, and (b) after connecting to the microgrid.

voltage balancer neither absorbs nor injects power to the L1N or L2N bus. The total power of 14 kW is supplied by the inverter.

By examination of the converters' transfer function matrices before and after connecting to the bipolar DC microgrid, the interactions among the control loops of the converters can be quantified. Considering Load#3, the transfer function relating the output current (load current) to the output voltage ($\triangle v_{o3}/\triangle i_{o3}$) is given by $T_{Load3}(1,2)$ before the converter is connected to the bipolar DC microgrid. The pole-zero map of $T_{Load3}(1,2)$ is shown in Figure 8.11a, which represents a stable system. Note that the converter is operating isolated from the microgrid. However, when it is connected to the microgrid, the interactions among the control loops of other converters impose undesired dynamics to the control system of Load#3. This is computed by (8.12), $T_{MG}(3,3)$, which represents an unstable system with right-hand poles as shown in Figure 8.11b. Hence, the additional poles and zeros are due to the interactions among the control systems of Load#3 and other converters in the microgrid.

Hence, if the dynamics of the input DC voltage source are ignored, the control system of Load#3 is stable; however, if the dynamics are included, the control system of the converter becomes unstable.

8.5.2   Mutual Interactions

This section explores the following: (1) how a load change on L1N bus influences the dynamics of the converters control systems across L2N bus; (2) how the Geršgorin bands can be utilized to identify the degree of coupling among various input–output pairs as the bipolar DC microgrid becomes more unbalanced; and finally (3) how the SVD analysis can predict the frequency of the unstable poles.

The mutual interactions among control systems of the converters connected to L1N and L2N terminals can be understood using (8.12). A change in the Load#3 (step change in i_{o3}) would have impacts not only locally on the dynamics of v_{o3}, but also on the dynamics of all other converters. These mutual interactions can be observed by the analysis of $\Delta v_{o2}/\Delta i_{o3}$ or $\Delta v_{o3}/\Delta i_{o2}$.

In this discussion, the rated power of Load#3 is increasing in three steps of 4 kW, 6 kW, and 8 kW, while the nominal power of Load#2 is constant at 2 kW. The small-signal dynamics of Δv_{o2} for step changes in Δi_{o3} are investigated from step response analysis of $T_{MG}(2,3)$, as it is presented in Figure 8.12a. The impact of Δi_{o3} on the dynamics of Δv_{o2} is rather insignificant, whereas Δi_{o2} affects the dynamics of

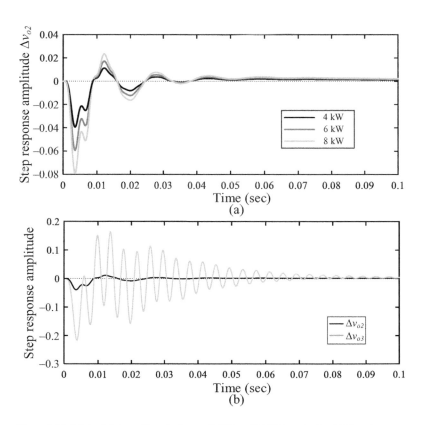

Figure 8.12 Mutual interactions between L1N and L2N terminals: (a) the step response of $\Delta v_{o2}/\Delta i_{o3}$ for different operating points of Load#3, and (b) the step response of $\Delta v_{o2}/\Delta i_{o3}$ and $\Delta v_{o3}/\Delta i_{o2}$ for a similar operating point.

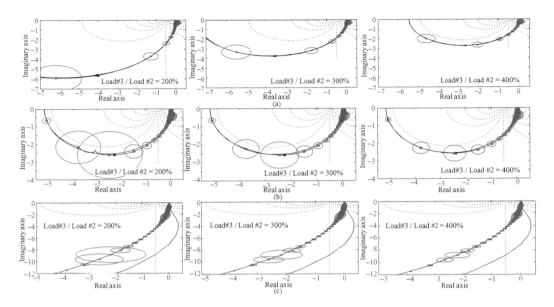

Figure 8.13 Geršgorin circles indicating the degree of coupling in $T_{MG}(s)$: (a) $T_{MG}(1,1)$, (b) $T_{MG}(2,2)$, and (c) $T_{MG}(3,3)$.

$\triangle v_{o3}$ quite significantly. A comparison between the step responses of $\triangle v_{o2}/\triangle i_{o3}$ and $\triangle v_{o3}/\triangle i_{o2}$ is illustrated in Figure 8.12b, which shows that the control system of Load#3 is more liable to disturbances.

The interactions among the control systems of converters connected to L1N and L2N terminals can be further investigated by applying the Geršgorin circle theorem to the transfer function matrix T_{MG}. The Geršgorin bands for diagonal elements of T_{MG}, including $T_{MG}(1,1)$, $T_{MG}(2,2)$, and $T_{MG}(3,3)$, are shown, respectively, in Figure 8.13a–c. The Geršgorin bands are calculated for three loading conditions of the bipolar DC microgrid. First, the unbalance ratio (defined as $Load\#3/Load\#2$) is 200%. Then this ratio is increased to 300% and further to 400%. As it is shown in Figure 8.13, with the increase of the unbalance ratio, the coupling among input–output pairs in the MIMO system becomes weaker, that is, the Geršgorin circles become thinner. Thus, in case the bipolar DC microgrid is more unbalanced, the overall MIMO system given by (8.12) is more diagonal, which means that $\triangle v_{o1}$ is weakly affected by changes in $\triangle i_{o2}$, $\triangle i_{o3}$, and $\triangle v_d$. This holds for the other two outputs as well.

Another useful linear tool to study the mutual interactions among control systems is the frequency-dependent RGA analysis, which is applied to $T_{MG}(jw)$ at various frequencies. The RGA values of $\triangle v_{o3}$ for four inputs and at the desired frequency range are presented in Figure 8.14a. Ideally, for a system with a low level of mutual interactions, the control system of the converter is only affected by its local disturbance. For instance, the ideal case for $\triangle v_{o3}$ is that RGA values at all frequencies are equal to one for $\triangle i_{o3}$, whereas they are zero for all other inputs ($\triangle i_{o2}$, $\triangle i_{o1}$, and $\triangle v_d$). However, as it is illustrated in Figure 8.14a, the control loop of $\triangle v_{o3}$ strongly interacts

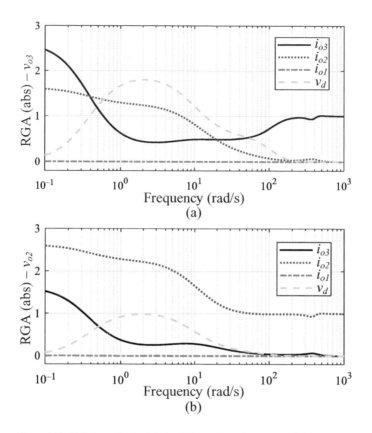

Figure 8.14 RGA analysis: (a) the RGA values for v_{o3}, and (b) the RGA values for v_{o2}.

with Δi_{o2} and Δv_d for the frequency range of 1–100 rad/s. This reveals that the voltage control loops of the converters connected to L1N bus are considerably affected by the load current dynamics of the converters connected to L2N bus. In case of Δv_{o2}, the dynamics of the voltage control loop are mainly influenced by the local disturbance Δi_{o2}, as it is indicated in Figure 8.14b. For all frequencies, the highest RGA values belong to Δi_{o2}.

The next step is to investigate the interactions among a vector of inputs and outputs. Using SVD analysis, the maximum amplification on the vector of outputs can be identified. Applying SVD analysis to $T_{MG}(s)$, the region of the singular values of $T_{MG}(jw)$, which is confined by $\overline{\sigma}_i$ and $\underline{\sigma}_i$, is presented in Figure 8.15a. As can be seen, for higher frequencies, if excited, the larger amplification of inputs would appear in the outputs. Moreover, referring to Figure 8.15a, $\overline{\sigma}_i$ and $\underline{\sigma}_i$ begin to change abruptly around frequencies between 1,400 and 1,800 rad/s. To better illustrate it, a condition number that is the ratio of $\overline{\sigma}_i$ / $\underline{\sigma}_i$ is defined.

For three unbalance ratios (200%, 300%, and 400%), the condition number for the frequency range of interest is plotted in Figure 8.15b. For all unbalance ratios, the minimum condition number is in the frequency range of 1,400 and 1,800 rad/s. The frequency of the unstable poles of $T_{MG}(jw)$, which is shown in Figure 8.15c, is

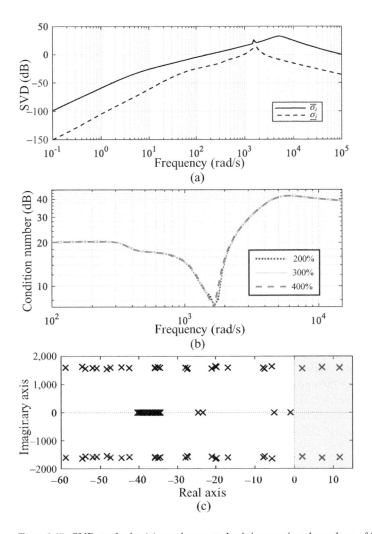

Figure 8.15 SVD analysis: (a) maximum and minimum singular values of $T_{MG}(jw)$, (b) condition number ($\overline{\sigma}_i / \underline{\sigma}_i$) for various unbalance ratios, and (c) the dominant poles of microgrid.

also within this frequency range. Therefore, the condition number obtained from SVD analysis can be regarded as a criterion to estimate the frequency of unstable poles. It does not provide information about whether the system is stable or not; instead, it predicts the frequency of unstable poles.

References

[1] L. Meng, Q. Shafiee, G. Ferrari Trecate, H. Karimi, D. Fulwani, X. Lu, and J. Guerrero, "Review on Control of DC Microgrids and Multiple Microgrid Clusters," *IEEE Journal of Emerging and Selected Topics in Power Electronics*, vol. 5, no. 3, pp. 928–948, 2017.

[2] A. Elsayed, A. Mohamed, and O. A. Mohammed, "DC Microgrids and Distribution Systems: An overview," *Electric Power Systems Research*, vol. 119, pp. 407–417, 2015.

[3] S. Anand, G. Fernandes, and J. Guerrero, "Distributed Control to Ensure Proportional Load Sharing and Improve Voltage Regulation in Low-Voltage DC Microgrids," *IEEE Transactions on Power Electronics*, vol. 28, pp. 1900–1912, 2013.

[4] J. Guerrero, J. Vasquez, J. Matas, L. de Vicuna, and M. Castilla, "Hierarchical Control of Droop-Controlled AC and DC Microgrids: A General Approach toward Standardization," *IEEE Transactions on Industrial Electronics*, vol. 58, pp. 158–172, 2011.

[5] H. Kakigano, Y. Miura, and T. Ise, "Low-Voltage Bipolar-Type DC Microgrid for Super High Quality Distribution," *IEEE Transactions on Power Electronics*, vol. 25, pp. 3066–3075, 2010.

[6] S. Tavakoli, J. Khajesalehi, M. Hamzeh, and K. Sheshyekani, "Decentralised Voltage Balancing in Bipolar DC Microgrids Equipped with Trans-Z-Source Interlinking Converter," *IET, Renewable Power Generation*, vol. 10, pp. 703–712, 2016.

[7] M. Ferrera, S. Litran, E. Duran Aranda, and J. Andujar Marquez, "A Converter for Bipolar DC Link Based on SEPIC-Cuk Combination," *IEEE Transactions on Power Electronics*, vol. 30, pp. 6483–6487, 2015.

[8] J. Lago and M. Heldwein, "Operation and Control-Oriented Modeling of a Power Converter for Current Balancing and Stability Improvement of DC Active Distribution Networks," *IEEE Transactions on Power Electronics*, vol. 26, pp. 877–885, 2011.

[9] E. Trinklein, G. Parker, D. Robinett, and W. A. Weaver, "Toward Online Optimal Power Flow of a Networked DC Microgrid System," *IEEE Journal of Emerging and Selected Topics in Power Electronics*, vol. 5, no. 3, pp. 949–959, 2017.

[10] S. D. Tavakoli, G. Kadkhodaei, M. Mahdavyfakhr, M. Hamzeh, and K. Sheshyekani, "Interlinking Converters in Application of Bipolar DC Microgrids," in *2017 8th Power Electronics, Drive Systems & Technologies Conference (PEDSTC)*. IEEE, pp. 37–42, 2017.

Part III

Prospect

9 Software-Defined Distribution Network and Software-Defined Microgrids

9.1 Motivation

This chapter envisions how power distribution grids in smart cities and their building blocks – networked microgrids – will evolve in the coming years.

More than 80% of the US population currently reside in cities, and this is also where 80% of the nation's energy is consumed [1]. At the heart of each city's infrastructure is its urban power distribution network (UDN), which supports urban systems such as energy, food, transportation, communication, and many others that are key to citizens' living and working lives [2]. In smart cities, UDNs should be resilient, scalable, and flexible so that they can meet the ever-increasing needs of people in urban environments. Existing UDNs, however, were not designed for smart cities and cannot sustain their growth [3].

First, UDNs have been severely impacted by the ubiquitous integration of distributed energy resources (DERs), most of which are photovoltaic (PV) units. For example, today, more than 15,000 residential solar PV projects have been installed in the state of Connecticut alone, and more than 4,600 additional projects are currently in progress. This number is projected to be quadrupled within the next four years. In 2015, a new PV was tied to the distribution grid every two minutes in the United States [4], and this pace will likely increase due to drops in PV costs [5] (down to $0.06/kWh in 2020 without incentives, predicted by Department of Energy). Because of the intermittency and fast ramping of PV, however, urban grid infrastructures and customers are suffering from voltage and frequency fluctuations. To maintain power quality and reliability, grid devices such as transformer taps are being forced to adjust frequently, which has led them to rapidly reach their end of lives or suffer from premature failures. As an example, Hawaii utilities reported that their onload tap changer (OLTC) transformers, traditionally maintenance-free during a 40-year lifespan, had to be maintained every three months and retired within two years because PV-induced voltage fluctuations made the OLTC adjust more than 300 times per day [6]. Further, the lifespan of battery storage to smooth PV output could be reduced from 15 years to less than five years when the PV penetration level reaches 50% [7]. This will place an unprecedented burden on stakeholders and citizens.

The second challenge faced by UDNs is the frequent blackouts caused by extreme weather events. For instance, in 2012, Hurricane Sandy left millions of people in New York City without power for days (some regions lost power for up to two months).

Reports from the National Oceanic and Atmospheric Administration (NOAA) indicate a trend that extreme weather events are likely to increase as the climate continues to change [3]. Proactive actions therefore must be taken to make UDNs more resilient to extreme weather conditions.

Addressing the first challenge, that is, the smooth integration of renewables, calls for innovative techniques for the real-time operation optimization of large UDNs [8–12] such that distributed generation and other controllable devices are economically operated with smooth grid voltages. For the second challenge, a strong consensus across academia, industry, and government is that microgrids can provide an emerging and promising paradigm for urban electricity [13–19]. Because of the concentration of population and therefore critical load within urban cores, microgrids are being increasingly deployed. A smart city zone is expected to have many microgrids operated by various stakeholders. Networking local microgrids and allowing them to support coordinately various smart city functions and architectures will provide a potent resolution to the second challenge, viz. increasing resilience during extreme weather events. However, both solutions share a common bottleneck: a scalable and high-speed communication and computing infrastructure that does not yet exist.

This scalable and high-speed cyberinfrastructure is of critical importance for enabling sustainable and resilient urban power distribution systems because both optimizing operations of UDNs and networking microgrids rely on collecting and processing data from various grid nodes in real time at different levels. Currently, UDNs are so sparsely sensed that they typically cannot be observed or controlled beyond distribution substations. Because traditional UDNs tend to have very limited bandwidths, it is impossible for them to receive information from massive new types of sensors (e.g., PV inverters, advanced metering infrastructure [AMI] meters, remote terminal units [RTUs] and microphasor measurement units [microPMUs]) with reasonable response times. The number of these active nodes will grow from tens to thousands and even millions in future UDNs. For example, Hartford, the capital of Connecticut, has over 120 distribution feeders. A typical feeder with 20% PV penetration may have about 1,000 load points and 200 PV units in the near future. This would result in more than 140,000 active nodes if every PV or load point had local control. Among those nodes, an AMI meter for a load point usually collects electricity usage data every 15 minutes, while a PV data acquisition unit collects various types of data (e.g., power production, inverter output and status, AC-side current and voltage) every few seconds to five minutes. RTUs will transmit distribution substation data every four seconds. Further, we will see wide adoptions of microPMUs, which capture network states at 512 samples per cycle (about 30,000 data packets per second). The speed of existing UDN communications, such as VHF and UHF narrow bandwidth radio or programmable logic controller (PLC), is in a range of kilobytes per second to megabytes per second. It is clear that the state-of-the-practice communication and information processing architecture of traditional UDNs will not be able to handle the gigantic volumes of data generated from hundreds of thousands of active nodes [12].

From the perspective of UDN management, existing centralized optimization techniques [20, 21] suitable for handling small-scale, passive UDNs will no longer be able to provide much-needed support for the future's large, active UDNs. Distributed algorithms based on decomposition and coordination, on the other hand, have been found to be promising for large-scale applications [22, 23]. However, the lack of a scalable and resilient gigabit UDN infrastructure has significantly hindered the development and implementation of the distributed algorithms needed for integrating massive renewable energy sources and ensuring electricity resilience in smart cities.

9.2 Software-Defined Distribution Network and Software-Defined Networked Microgrids

Motivated by the challenges detailed in the preceding section, we envision a game-changing way for distributing power: the software-defined distribution network (SD^2N), a novel gigabit urban infrastructure that integrates software-defined networking (SDN), real-time computing, Internet of Things (IoT) techniques, and distributed control and optimization algorithms for UDNs. In particular, our approach will address three critical issues and distribute algorithms among global or local computing resources to be connected via a resilient programmable communication network. The three breakthroughs are as follows:

- SD^2N architecture
  A scalable, self-reconfigurable, plug-and-play platform to coordinate the flows of power/data to and from large UDNs with ubiquitous integration of intermittent renewable energy sources.
- Distributed algorithms
  SD^2N-enabled distributed optimization and control algorithms that mitigate the impact of the high penetration of renewable energy resources while maximizing system efficiency under normal operations. This requires incorporating current network states and forward-looking information into a distributed optimization/control process.
- Software-defined networked microgrids (SDNM)
  SD^2N-enabled resilient networked microgrids. SDNM are capable of regulating microgrid interactions and maximizing electricity resilience under extreme events.

In order to build an SD^2N platform, the following issues must be investigated:

- How to leverage SDN and IoT techniques to design an SD^2N architecture to enable real-time data streaming, processing, storage, and feedback while tackling the stringent data availability and multilatency requirements in managing large UDNs.
- How to design and implement global/local computing infrastructures in SD^2N to unlock the potential of distributed algorithms for the two SD^2N applications.

The new software-defined networking and computing framework will be the first attempt to study how heterogeneous active nodes (e.g., PV inverters, microgrids, microPMUs) can communicate with each other cooperatively to achieve system-level objectives. SD^2N will offer insights and guidance in deploying gigabit networks to various large cyberphysical systems where monitoring and control decisions need to be made under critical time constraints based on distributed sensing, computing, and optimization. The SD^2N-enabled applications will be advanced beyond the present capability of legacy distribution management systems used in distribution control centers, injecting the utmost flexibility, agility and operational efficiency into UDNs. It will enable a truly scalable, self-adaptive, self-reconfigurable, and replicable power infrastructure that allows progressively deeper integration of DERs and is resilient to natural and manmade attacks.

The envisioned SD^2N architecture is illustrated in Figure 9.1. It contains two layers: the physical layer and the cyber layer, shown in the lower and upper parts of Figure 9.1, respectively. The physical layer contains a set of distribution substations, feeders, and microgrids connected to feeders. A distribution substation is a system that connects the high-voltage subtransmission system and the lower-voltage distribution feeders through distribution transformers and buses. It also undertakes the critical responsibilities of continuously measuring, monitoring, protecting, and controlling its section of the grid. A feeder is a power line or cable system that supplies power to load points such as buildings or industrial plants. Microgrids and networked microgrids are usually deployed at the edge of the UDN and are widely envisioned as building blocks of smart cities due to their potential to reduce a city's carbon footprint and enhance electricity resilience both under extreme events and on a daily basis [24–29].

The cyber layer contains the communication and computing infrastructure, which includes both local and wide-area communication networks, as well as local and global computing resources. Inside a substation, an IoT-enabled hybrid communication fabric is used for real-time data transport, and a real-time data analytics platform (deployed on the local computing resource) is used for computation and storage to support various optimization and management tasks. The hybrid communication fabric includes real-time edge networks and an ultrafast programmable network backbone. The edge networks are connected to various substation resources through IoT-enabled lightweight data connectors for distributed data acquisition. These real-time data measurements are streamed into the data analytics platform for real-time monitoring, optimization, and management of the substation. Each substation is connected to the subtransmission grid at the physical layer and is also connected to a distribution operation center (via a wide-area network) at the cyber layer. The communication and computing infrastructure for a microgrid is similar to that for a substation. Under normal conditions, a microgrid is connected to a feeder and is controlled by a microgrid central controller (MGCC), which further communicates with the distribution operation center indirectly via the corresponding substation. During emergency conditions (e.g., a blackout of the main grid), microgrids will be disconnected from the main UDN, and several microgrids with close electrical or spatial proximity can coordinate with each other as networked microgrids to perform an emergency response.

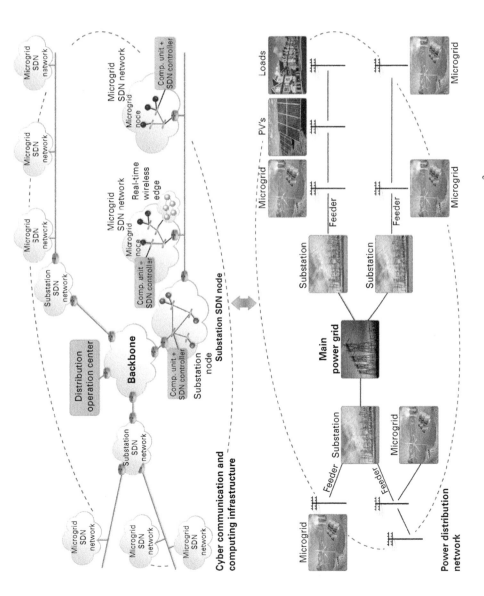

Figure 9.1 Architectural overview of the software defined distribution network (SD^{2}N). Figure courtesy of Yelena Belyavina, Dr. Meng Yue, Dr. Song Han, Dr. Bing Wang, and Dr. Peter B. Luh [2]

9.3   Scalable and Resilient Network Management

9.3.1   SDN-Enabled Communication Infrastructure

The SD^2N architecture for large-scale distribution networks contains two components: an SDN-based communication infrastructure and a distributed and scalable real-time data analytics platform. The integrated communication and computing framework will significantly improve the scalability, reliability, and real-time performance of routine and emergency distribution/substation/microgrid operations.

A key advantage of SDN is that it allows programmable access to the network switches in real time, which provides unprecedented flexibility and ease in managing large-scale computer networks. Using SDN as the communication infrastructure for power systems is advantageous for several reasons. First, SDN allows the network to be reconfigured dynamically in real time so that it can respond to failures, congestion, or cyberattacks. This will allow for significantly more resilient communication support to various applications in the power grid. Secondly, SDN provides flexible functions to support the diverse quality of service (QoS) requirements for data packets in power systems, where some are small control packets that require latency in milliseconds while others can tolerate much larger latencies in seconds or minutes. In addition, SDN adopts open protocols in network switches and supervisory controllers, which makes it much easier to develop new applications and enable fast innovation in power systems.

The author's team has designed a novel architecture and a set of customized techniques that employ SDN as the communication infrastructure for a single microgrid. We are extending the design significantly to large-scale distribution networks, as illustrated in Figure 9.2. The proposed system design uses a hierarchical architecture, where on the upper level, an SDN controller (or multiple SDN controllers for reliability and scalability) manages the gateway switches in a set of microgrids and substations; on the lower level, an SDN controller manages the routers inside an individual microgrid or substation. Some devices (e.g., RTUs, microPMUs, PLCs) are directly connected to SDN switches and hence are directly connected to the programmable network. Other devices (e.g., IoT sensors) may be indirectly connected to the programmable network through a gateway. Three new features will be built in the SDN-enabled communication architecture:

- Ensuring fast fault detection and route reconfiguration and satisfying QoS requirements in the large-scale network scenario, which may require the coordination of upper- and lower-level SDN controllers.
- Automatically configuring the communication network. This involves building a self-configurable platform using SDN to manage dynamic addition and deletion of devices as well as scalable and reliable multicast communications when applications have diverse QoS requirements.
- Achieving scalable and real-time monitoring of network conditions, which is critical for obtaining a global view of the network for network management. In addition, to deal with the heterogeneous communication protocols currently used by the

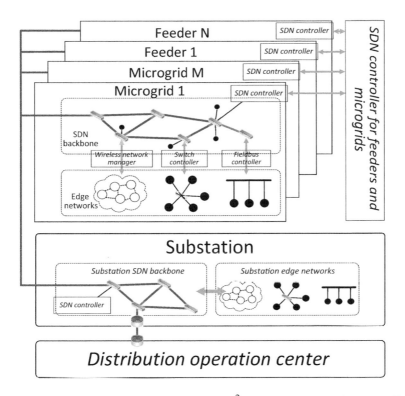

Figure 9.2 The networking architecture in SD^2N. Figure courtesy of Dr. Song Han, Dr. Bing Wang, and Dr. Peter B. Luh [2]

communication devices in distribution networks, the conversion of the heterogeneous protocols into a uniform format is needed.

9.3.2 Scalable and Distributed Real-Time Data Analytics Platform for SD^2N

This platform (see Figure 9.3) is designed to perform the following functions: digest the high volume of real-time measurements gathered from the distribution network, support continuous data analytics and real-time monitoring and optimization at various levels, and provide timely responses to the system and operators. It will be deployed on the computing resources in the microgrids, substations, and the control center. Unified streaming application program interfaces (APIs) will be created among these computing resources, which together form the scalable and distributed computing infrastructure for SD^2N. Real-time measurements collected from various data sources in the distribution network (from both the physical power network and the cyber communication network) will be streamed, stored, and processed in a bottom-up manner at different levels of the distributed computing infrastructure, which will support both multi-scale network management and power system optimization and control.

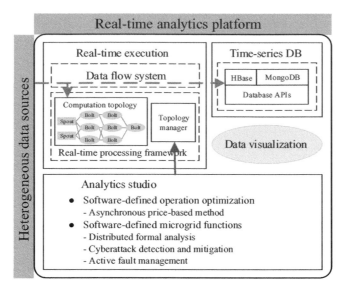

Figure 9.3 Overview of the distributed real-time data analytics platform.

The following key components differentiate the SD^2N data analytics platform from existing general-purpose computing platforms:

- A unified messaging protocol to support massive heterogeneous real-time data streams
- A distributed time-series database specifically designed for distribution network monitoring, analytics, and optimization applications
- An analytics development studio for the design of data and control flows related to microgrid/substation management analytics and optimization tasks
- An elastic real-time parallel processing framework on which the developed models and analytics modules will be deployed to distribute the computation in a hierarchical computing infrastructure

The SD^2N data analytics platform can be connected to various distribution network resources through lightweight data connectors for distributed data acquisition.

9.4   Distributed Advanced Energy Management System

SD^2N transforms the infrastructure layer, as well as the communications and data layers, into unprecedentedly flexible and resilient services. As a result, SD^2N will allow for the development of a set of innovative grid analytics needed for operating real-time distribution systems and networked microgrids so that they can be implemented in a distributed manner, as needed, for the scalability of the proposed solution.

9.4.1 SD^{2}N-Enabled Distributed Distribution System State Estimation

Topology identification is the first step in state estimation. Traditional topology identification methods, however, no longer work because of (1) the huge dimensions of the distribution grid compared to traditional transmission networks, (2) unobservability of massive underground branches, and (3) the coexistence of radial and meshed topologies subject to reconfigurations. Further, no existing distribution system state estimation (DSSE) methods are tractable for the real-time operation of million-node distribution grids. On the one hand, the number of measurements (M) is much less than the number of system states (N), making the existing DSSE (e.g., weighted least square and its variants) no longer viable. On the other hand, M itself is increasingly large, and real-time/quasi-real-time measurement data from massive sensing devices pose a big data challenge that centralized DSSE methods cannot tractably handle. For instance, high velocity streams from microPMUs at 512 samples per second can accumulate gigabytes of data per day for a single microgrid.

The SD^{2}N architecture and grid edge resources offer new opportunities for novel distributed DSSE (D-DSSE) solutions to address the challenges of both sparse measurements and big data issues. Scalability is built into the integration of machine learning [29, 30], compressive sensing (CS) [31], and distributed state estimation [32] using the distributed computing resources in SD^{2}N.

9.4.2 SD^{2}N-Enabled Distribution Optimal Power Flow

The OPF considering Volt/Var, loss minimization, minimum control actions (e.g., for load tap changers and capacity banks with discrete decisions to minimize voltage violations and prolong device lifetimes), and minimum greenhouse gas emissions can be innovatively formulated as a mixed-integer multiobjective economic dispatch problem. The multiple objectives of the preceding formulation will first be linearly combined. The resulting mixed-integer linear programming (MILP) problem is combinatorial and will be solved by using a synergistic integration of the latest Surrogate Augmented Lagrangian Relaxation and branch-and-cut [33, 34]. With weights varying at small increments, the Pareto frontier can be obtained for system operators to select weights judiciously. The method will then be made distributive and asynchronous for innovative price-based real-time coordinations of a very large number of distributed devices using the Lyapunov stability concept. The intermittency of PV generation will be managed by using an inventive integration of Markovian and interval modeling approaches to avoid the complexity of or the lack of representativeness of scenario-based approaches [35, 36].

9.4.3 Resilience Engineering for Future Power Networks

The Presidential Policy Directive (PPD-21) [37] defines resilience as "the ability to prepare for and adapt to changing conditions and withstand and recover rapidly from disruptions." There is a strong consensus across the community of academia [38–40], industry [13, 14, 39], and government [41, 42] that the concept of resilience will play a major role in assessing the extent to which critical infrastructures are prepared to deal with the threats they face [16]. In the US, significant blackouts have risen from 76 in 2007 to 307 in 2011, leading to enormous economic and social upheavals. In 2012, Hurricane Sandy left more than 8 million customers without electricity in 10 U.S. states and the District of Columbia. In addition to a death toll of 113, it was reported that the total losses from Hurricane Sandy exceeded \$50 billion. The possibility of more frequent natural and manmade events, especially weather extremes, suggests that the rate of forced outages and bunching failures in power systems will only increase over time. As a result, monitoring and managing power system resilience has been a focus area of increasing emphasis for the development of future smart grids.

The Electric Power Research Institute (EPRI) recently summarized [42] utility hardening efforts for enhancing power distribution resilience. However, research on power system resilience is still at an early stage and entails significant challenges [16]. First of all, one needs to create working metrics that can be used to characterize and provide information on the resiliency of power infrastructures [43, 44]. Few quantitative resilience metrics have been developed for power grids thus far [45]. Further, there is a need to develop model-based and/or data-driven methods for effectively computing the metrics. In fact, a US Department of Energy (DOE) workshop for grid resilience has identified this as the top research gap to be filled [46]. Specifically, the priority issues [46] are to build (i) predictive models to quantify grid resilience to threats and load changes through the use real-time data; (ii) technologies for rapid, proactive damage assessment to facilitate rapid recovery under extreme events; and (iii) decision support to determine resilience enhancement priorities. The design of those methods faces stringent challenges: (i) Scalability. Resilience is of greatest concern during the low-probability events with high consequences that cannot be captured by the common $N - x$ security analyses or by the reliability evaluation for averaged performance. For these cases, traditional approaches such as the Monte Carlo simulations become computationally intractable for predicting resilience at a system level. (ii) Dependability. Events of concern are disruptions or acute events that occur over a relatively short time period. The approaches should be highly reliable because each event might have big impacts. It is challenging to ensure robust and high-fidelity solutions under dynamic, stochastic conditions and uncertainties. (iii) Intelligence. The algorithms should be able to infer from big data and run with little or no supervision.

SD^2N, if successfully built, can establish a platform for addressing the challenges of electricity resilience. Meanwhile, major efforts still need to be taken to establish a resilience engineering framework for future grids and to develop scalable cybertools for predicting and managing grid resilience. Inspired by resiliency theory for complex systems [47], a regime-shift approach may be pursued to gain a deeper understanding of power system resilience. Generally, when a system undergoes a change from one characteristic pattern or set of behaviors to another, the change is called a regime shift [48]. Power system resilience can then be understood as the tendency of a power

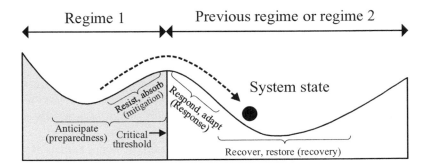

Figure 9.4 Resilience from a regime-shift perspective.

system subject to change to remain within a regime, continually changing and adapting yet remaining within critical thresholds; it also reflects the capacity to cross thresholds into a new desirable regime after a catastrophic shift. Figure 9.4 illustrates the alternative definition, where the cups or valleys represent regimes [49] in which power systems can function and be structured.

The regime-shift definition is consistent with existing definitions, covering both "before adverse event" and "after adverse event" components. The advantage of using this definition is that the well-laid foundation of regime shift theory may open the door for creating operational, rigorous metrics and methods for the forecasting of power system resilience. Resilience can be interpreted as the perturbation magnitude a system can absorb without experiencing cartographic regime shifts. Thus, the following quantities can serve as basic resilience metrics that may lead to specific resilience metrics for different applications:

- *"Spatial" metric*: the distance between the current state and the critical threshold of regime shift
- *Time metric*: the time it takes the system to reach the critical threshold
- *Transition metric*: the probability of regime shift
- *Recovery metric*: the time needed to bring the system back to a desirable regime if a catastrophic shift does occur

The preceding metrics provide a guiding vision. For power distribution and networked microgrids, it is still an open problem to create computational cybertools that help power engineers to (1) monitor power system resilience by the forecasting of the occurrence of regime shifts in large-scale distribution systems and networked microgrids while (2) providing guidance on regime-shift management.

References

[1] L. Parshall, K. Gurney, S. A. Hammer, D. Mendoza, Y. Zhou, and S. Geethakumar, "Modeling Energy Consumption and CO_2 Emissions at the Urban Scale: Methodological Challenges and Insights from the United States," *Energy Policy*, vol. 38, no. 9, pp. 4765–4782, 2010.

[2] P. Zhang, B. Wang, P. B. Luh, and S. Han, *US Ignite: Focus Area 1: SD^2N: Software-Defined Urban Distribution Network for Smart Cities*, Grant Proposal for CNS-1647209, National Science Foundation, 2016.

[3] Electricity Advisory Committee, "Keeping the Lights on in a New World," US Department of Energy, Washington, DC, vol. 50, 2009.

[4] Eversource Energy Center, "Electric Grid Hardening: Integrating Renewable Generation," www.eversource.uconn.edu/electric-grid-hardening/integrating-renewable-generation, May 2016.

[5] Office of Energy Efficiency and Renewable Energy, "Enabling Extreme Real-Time Grid-Integration of Solar Energy (ENERGISE)," Department of Energy DE-FOA-0001495, 2016.

[6] T. Sokugawa and M. Shawver, "Big Data for Renewable Integration at Hawaiian Electric Utilities," presentation at the DistribuTech Conference and Exhibition, Orlando, FL, February 9–11, 2016.

[7] P. Du and N. Lu, *Energy Storage for Smart Grids Planning and Optimization for Renewable and Variable Energy Resources.* Elsevier, 2015.

[8] E. Provata, D. Kolokotsa, S. Papantoniou, M. Pietrini, A. Giovannelli, and G. Romiti, "Development of Optimization Algorithms for the Leaf Community Microgrid," *Renewable Energy*, vol. 74, pp. 782–795, 2015.

[9] M. B. Shadmand and R. S. Balog, "Multi-Objective Optimization and Design of Photovoltaic-Wind Hybrid System for Community Smart DC Microgrid," *IEEE Transactions on Smart Grid*, vol. 5, no. 5, pp. 2635–2643, 2014.

[10] X. Ma, Y. Wang, and J. Qin, "Generic Model of a Community-Based Microgrid Integrating Wind Turbines, Photovoltaics and CHP Generations," *Applied Energy*, vol. 112, pp. 1475–1482, 2013.

[11] L. Che, X. Zhang, M. Shahidehpour, A. Alabdulwahab, and A. Abusorrah, "Optimal Interconnection Planning of Community Microgrids with Renewable Energy Sources," *IEEE Transactions on Smart Grid*, vol. 8, no. 3, pp. 1054–1063, 2017.

[12] D.-M. Bui, C.-H. Wu, S.-L. Chen, C.-H. Huang, and K.-Y. Lien, "Installation on a Community-Sized DC Microgrid System and Suggestion on a DC Short-Circuit Test Procedure," in *2014 China International Conference on Electricity Distribution (CICED).* IEEE, pp. 1819–1825, 2014.

[13] A. Rose, "Economic Resilience and Its Contribution to the Sustainability of Cities," in *Resilience and Sustainability in Relation to Natural Disasters: A Challenge for Future Cities.* Springer, pp. 1–11, 2014.

[14] L. Molyneaux, L. Wagner, C. Froome, and J. Foster, "Resilience and Electricity Systems: A Comparative Analysis," *Energy Policy*, vol. 47, pp. 188–201, 2012.

[15] J. Inglis, S. Whittaker, A. Dimitriadis, and S. Pillora, "Climate Adaptation Manual for Local Government: Embedding Resilience to Climate Change," Australian Centre of Excellence for Local Government, University of Technology, Sydney. https://opus.lib.uts.edu.au/handle/10453/42120 2014.

[16] J. Carlson, R. Haffenden, G. Bassett, W. Buehring, M. Collins III, S. Folga, F. Petit, J. Phillips, D. Verner, and R. Whitfield, "Resilience: Theory and Application." Argonne National Lab. (ANL), Argonne, IL (United States), Tech. Rep., 2012.

[17] P. Zhang, G. Li, and P. B. Luh, "Reliability Evaluation of Selective Hardening Options," State of Connecticut, Tech. Rep., May 2015.

[18] G. Li, P. Zhang, P. B. Luh, W. Li, Z. Bie, C. Serna, and Z. Zhao, "Risk Analysis for Distribution Systems in the Northeast US under Wind Storms," *IEEE Transactions on Power Systems*, vol. 29, no. 2, pp. 889–898, 2014.

[19] Z. Bie, P. Zhang, G. Li, B. Hua, M. Meehan, and X. Wang, "Reliability Evaluation of Active Distribution Systems Including Microgrids," *IEEE Transactions on Power Systems*, vol. 27, no. 4, pp. 2342–2350, 2012.

[20] Y. Riffonneau, S. Bacha, F. Barruel, S. Ploix, "Optimal Power Flow Management for Grid Connected PV Systems with Batteries," *IEEE Transactions on Sustainable Energy*, vol. 2, no. 3, pp. 309–320, 2011.

[21] S. Bruno, S. Lamonaca, G. Rotondo, U. Stecchi, and M. La Scala, "Unbalanced Three-Phase Optimal Power Flow for Smart Grids," *IEEE Transactions on Industrial Electronics*, vol. 58, no. 10, pp. 4504–4513, 2011.

[22] E. Dall'Anese, H. Zhu, and G. B. Giannakis, "Distributed Optimal Power Flow for Smart Microgrids," *IEEE Transactions on Smart Grid*, vol. 4, no. 3, pp. 1464–1475, 2013.

[23] T. Erseghe, "Distributed Optimal Power Flow Using ADMM," *IEEE Transactions on Power Systems*, vol. 29, no. 5, pp. 2370–2380, 2014.

[24] M. B. Shadmand, R. S. Balog, and M. D. Johnson, "Predicting Variability of High-Penetration Photovoltaic Systems in a Community Microgrid by Analyzing High-Temporal Rate Data," *IEEE Transactions on Sustainable Energy*, vol. 5, no. 4, pp. 1434–1442, 2014.

[25] L. Mariam, M. Basu, and M. F. Conlon, "Community Microgrid Based on Micro-Wind Generation System," in *2013 48th International Universities' Power Engineering Conference (UPEC)*, IEEE, pp. 1–6, 2013.

[26] L. Mariam, M. Basu, and M. F. Conlon, "Sustainability of Grid-Connected Community Microgrid Based on Microwind Generation System with Storage," in *2014 IEEE 23rd International Symposium on Industrial Electronics (ISIE)*, IEEE, 2014, pp. 2395–2400.

[27] L. Mariam, M. Basu, and M. F. Conlon, "Development of a Simulation Model for a Community Microgrid System," in *2014 49th International Universities Power Engineering Conference (UPEC), 2014 49th International Universities*, IEEE, pp. 1–6, 2014.

[28] X. Lu, S. Bahramirad, J. Wang, and C. Chen, "Bronzeville Community Microgrids: A Reliable, Resilient and Sustainable Solution for Integrated Energy Management with Distribution Systems," *Electricity Journal*, vol. 28, no. 10, pp. 29–42, 2015.

[29] S. Bahramirad, A. Khodaei, J. Svachula, and J. R. Aguero, "Building Resilient Integrated Grids: One Neighborhood at a Time," *IEEE Electrification Magazine*, vol. 3, no. 1, pp. 48–55, 2015.

[30] Y. Liao, Y. Weng, and R. Rajagopal, "Urban Distribution Grid Topology Reconstruction via Lasso," in *Power and Energy Society General Meeting (PESGM), 2016*, IEEE, pp. 1–5, 2016.

[31] S. S. Alam, B. Natarajan, and A. Pahwa, "Distribution Grid State Estimation from Compressed Measurements," *IEEE Transactions on Smart Grid*, vol. 5, no. 4, pp. 1631–1642, 2014.

[32] M. M. Nordman and M. Lehtonen, "Distributed Agent-Based State Estimation for Electrical Distribution Networks," *IEEE Transactions on Power Systems*, vol. 20, no. 2, pp. 652–658, 2005.

[33] M. A. Bragin, P. B. Luh, J. H. Yan, N. Yu, and G. A. Stern, "Convergence of the Surrogate Lagrangian Relaxation Method," *Journal of Optimization Theory and Applications*, vol. 164, no. 1, pp. 173–201, 2015.

[34] M. A. Bragin, P. B. Luh, J. H. Yan, and G. A. Stern, "Novel Exploitation of Convex Hull Invariance for Solving Unit Commitment by Using Surrogate Lagrangian Relaxation and Branch-and-Cut," in *2015 IEEE Power & Energy Society General Meeting*, IEEE, pp. 1–5, 2015.

[35] Y. Yu, P. B. Luh, E. Litvinov, T. Zheng, J. Zhao, and F. Zhao, "Grid Integration of Distributed Wind Generation: Hybrid Markovian and Interval Unit Commitment," *IEEE Transactions on Smart Grid*, vol. 6, no. 6, pp. 3061–3072, 2015.

[36] C. Wang, P. B.-S. Luh, and N. Navid, "Ramp Requirement Design for Reliable and Efficient Integration of Renewable Energy," *IEEE Transactions on Power Systems*, vol. 32, no. 1, pp. 562–571, 2017.

[37] Presidential Policy Directive, "Critical Infrastructure Security and Resilience," February 12, 2013.

[38] C. S. Holling, "Resilience and Stability of Ecological Systems," *Annual Review of Ecology and Systematics*, vol. 4, no. 1, pp. 1–23, 1973.

[39] S. Martin, G. Deffuant, and J. M. Calabrese, "Defining Resilience Mathematically: From Attractors to Viability," in *Viability and Resilience of Complex Systems*. G. Deffuant and N. Gilbert (eds.), Springer, pp. 15–36, 2011.

[40] C. Folke, S. R. Carpenter, B. Walker, M. Scheffer, T. Chapin, and J. Rockström, "Resilience Thinking: Integrating Resilience, Adaptability and Transformability," *Ecology and Society*, vol. 15, no. 4, https://www.ecologyandsociety.org/vol15/iss4/art20 2010.

[41] D. Stark, "On Resilience," *Social Sciences*, vol. 3, no. 1, pp. 60–70, 2014.

[42] M. McGranaghan, M. Olearczyk, and C. Gellings, "Enhancing Distribution Resiliency: Opportunities for Applying Innovative Technologies," *Electricity Today*, vol. 28, no. 1, pp. 46–48, 2013.

[43] R. J. Campbell, *Weather-Related Power Outages and Electric System Resiliency*, Washington: Congressional Research Service, Library of Congress, 2012.

[44] Q. Chen and L. Mili, "Composite Power System Vulnerability Evaluation to Cascading Failures Using Importance Sampling and Antithetic Variates," *IEEE Transactions on Power Systems*, vol. 28, no. 3, pp. 2321–2330, 2013.

[45] P. E. Roege, Z. A. Collier, J. Mancillas, J. A. McDonagh, and I. Linkov, "Metrics for Energy Resilience," *Energy Policy*, vol. 72, pp. 249–256, 2014.

[46] Office of Electricity Delivery and Energy Reliability Smart Grid R&D Program, "Draft Summary Report: 2014 DOE Resilient Electric Distribution Grid R&D Workshop," June 11, 2014.

[47] M. Scheffer, S. Carpenter, J. A. Foley, C. Folke, and B. Walker, "Catastrophic Shifts in Ecosystems," *Nature*, vol. 413, no. 6856, p. 591, 2001.

[48] H. Cabezas and T. Eason, "Fisher Information and Order," *San Luis Basin Sustainability Metrics Project: A Methodology for Assessing Regional Sustainability. US EPA Report: EPA/600/R-10/182*, pp. 163–222, 2010.

[49] M. Scheffer, Critical Transitions in Nature and Society (Princeton Studies in Complexity, 16) 2009.

10   Future Perspectives
Programmable Microgrids

What is the future of the power grid, the largest engineered system ever created by mankind? We believe autonomy will be the ultimate feature of future power infrastructures. In particular, microgrids and networked microgrids will play a key role in transforming today's community power infrastructures into tomorrow's self-configuring, self-healing, self-optimizing, and self-protecting power grids.

Although microgrids are effective and promising, transforming community power infrastructures into microgrids remains prohibitively difficult, which has hindered broader adoption of this technology. Six critical issues remain:

1. *Hardware dependence.* Forming a microgrid requires embedding protection, automation, and control (PAC) in expensive hardware facilities. Coping with the challenges posed by renewables and storms demands frequent, prohibitively expensive retrofitting and forced redesigns of PAC systems [1]. Moreover, to install extra dispatchable generators and PAC facilities requires a large designated space [2].
2. *Lack of a scalable, high-speed communication and computing infrastructure.* Current technology cannot process gigantic volumes of data [3] to support microgrid plug-and-play [4, 5] or network community microgrids [6].
3. *Limited and unscalable analytics.* The frequent changing of statuses, ubiquitous uncertainties, fast ramping, and nonsynchronism that characterize low-inertial community microgrids [7] creates intractable challenges to optimizing operations and reliably assessing and enhancing stability [8], all of which is key to microgrids' ability to serve as dependable resiliency resources [9].
4. *Cybersecurity and privacy* are of concern to the community [10–12].
5. *Lack of dispatchable generation* to enable seamless islanding [13–15].
6. *Social acceptance issues.* Research on the social acceptance of microgrids remains very limited, as the technology has only recently emerged [16, 17]. What factors affect social acceptance of microgrids, especially those that allow nearby stakeholders to connect to the microgrid as a "prosumer," remains an open question [18, 19].

10.1   Smart Programmable Microgrids

The key to solving these challenges lies in creating smart programmable microgrids (SPMs). This can only be achieved by pioneering new technologies that enable software-defined, hardware-independent community microgrid functions. The key innovation is a programmable platform that integrates software-defined networking (SDN), real-time computing, and the Internet of Things (IoT) to enable the following:

(a) Software-defined operation optimization
(b) The conversion of community facilities such as solar energy, storage, and waste management facilities into dispatchable resources
(c) Networking community microgrids without requiring expensive grid/communication infrastructure modifications.

The author's team will address the six aforementioned critical issues and will virtualize microgrid functions among global or local computing resources connected via a resilient programmable communication network. To achieve our main objective, we will contribute the following:

1. **Engineer a programmable microgrid** – a scalable, self-reconfigurable, plug-and-play platform to securely and efficiently manage the power/data flow to and from community microgrids. Our SPMs will virtualize traditionally hardware-dependent microgrid functions as flexible software services, fully resolving hardware dependence issues and enabling unprecedentedly low capital expenditure (CAPEX) and operating expense (OPEX).
2. **Pioneer a concept of "software-defined operation optimization" for microgrids**, where operation objectives, grid connections, and DER participation will be defined by software and plug-and-play so that they can be quickly reconfigured, based on the development of modularized and tightened models and a novel asynchronous price-based decomposition-and-coordination method.
3. **Devise software-defined distributed formal analysis (DFA)** for tractable stability assessment under heterogeneous uncertainties and the plug-and-play of microgrid components or microgrids.
4. **Establish software-defined security and privacy solutions** to protect microgrids from corrupted devices and other attacks and to ensure privacy and fairness to microgrid prosumers.
5. **Unlock the potential of community facilities**, such as converting anaerobic biomass digesters (AD) and wastewater treatment plants (WWTP) into environmentally friendly and dispatchable DERs.
6. **Understand what social, environmental, economic, institutional, and contract-specific factors shape whether residents support the development of SPMs** and how they vary based on proximity to the facility and ability to connect to SPMs as "prosumers."

10.2   Evaluation of Programmable Microgrids

An SPM testbed can be built by leveraging a cyberphysical networked microgrids testbed developed for enabling resilient microgrids/NMs through fast programmable networks, as supported by several projects mentioned in Chapter 1. Characteristics of the SPM testbed include the following:

1. Reconfigurable and distributed testbed architecture.
2. Federated real-time digital simulators connected with dozens of networked SDN switches, real-time computing servers, and IoT devices for large connected community grid studies.
3. Support of a wide spectrum of industrial communication protocols while interfacing with open source cyberlibraries for cyberattack studies.

Using our realistic-scale, high-fidelity testbed, we will execute the two stages of demonstration and evaluation in close collaboration with utilities and industry partners.

1. *Test SPM prototype.* In the first stage, we will test the SPM prototype in a virtual HIL environment that uses real-time measurements from test systems and closely mimics community microgrids. Specifically, the environment will contain real SDN communication and computing infrastructures, as well as protection and microgrid control systems virtualized in distributed computing systems. Meanwhile, the rest of the system will employ trace-driven simulation using real-time measurements (e.g., load curves, PV power) obtained from utilities. We will launch cyberattacks through the cyberlibraries in the network infrastructure and will simulate various SPM conditions and extreme events (e.g., extreme load and extreme weather).
2. *Evaluation of proposed SPM on community microgrids.* We will deploy the SPM functions on networked community microgrids and evaluate the following:
   (a) How SPMs can significantly reduce CAPEX and OPEX to improve social welfare.
   (b) How SPM improves hosting capacity with regard to renewables, electricity resilience, stability, power quality, and cyberphysical security using both deterministic and randomized tests.
   (c) How effectively and efficiently our SPM methods can convert waste management facilities into dispatchable, plug-and-play DERs.
   (d) How various factors influence the social acceptance of SPM to offer insights into the modernization of our nation's grid. Further, we will test the capability of SPM to network neighboring microgrids in order to maximize electricity resiliency and community connectivity.

We expect our SPM to achieve unprecedented cost-effectiveness, resilience, and social benefits for community customers. We will verify the performance of SPM under real-time constraints and various scenarios through key performance indicators summarized in Figure 10.1.

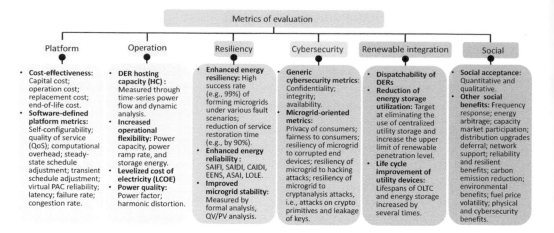

Figure 10.1 Metrics of evaluation for SPM.

10.3 Beyond Resilience

Future smart grids will employ ultrafast programmable networks as their core communication infrastructure. Because these networks can provide time delay guarantees, automatic failure recovery, and communication speed control, they have the potential to transform power grids into autonomic networks and flexible services that will be resilient to grid changes, DER injections, faults, and disastrous events.

A fully programmable smart grid will leverage the IoT and real-time computing technologies to enable real-time data streaming, processing, storage, and decision making while tackling the stringent data availability and multilatency requirements of managing smart grids. Scalable tools and distributed algorithms form the top layer of the architecture to facilitate the development and validation of advanced grid analytics, which will then virtualize the hardware-dependent PAC functions as software services. The combination of data-science-driven and model-based methods will enable power systems to become self-modeling [20] and self-optimizing autonomic machines. With this design, the future programmable power grid has the potential to transform the infrastructure layer as well as the communication and control layers of smart grids into unprecedentedly flexible and resilient services, which will then enable an enhanced system layer to seamlessly integrate the high penetration of renewable energy resources into the grids. Because this integration will enable unprecedented resilience in a cost-effective, secure, and reliable manner, it will also create opportunities to resolve many long-standing intractable problems in power and energy grids.

For decades, *reliability* has been a dominating metric for measuring electric power system performance [21, 22]. Given the advent of stronger weather events and an increasing number of cyberattacks in recent years, reliable safeguards and technologies are more necessary than ever so that this very important utility can be protected. Accordingly, *resilience* is emerging as a new metric for power system performance;

Table 10.1 Differences between renewability and resilience.

Renewability/regenerability	Resilience
Proactiveness: actively extending and enabling capacity	Passivity: built-in capacity
Noninfrastructure solutions: "Crowdsourced" solutions from ubiquitous edge resources	Infrastructure-reliance solutions
Inclusiveness: enabling DERs as reliable resilience resources	Exclusiveness: prudent to deep integration of DERs due to potential negative impact during emergencies
Aggregation: enabling DER aggregation for providing grid flexibility	Distribution: using fast ramping resources to compensate intermittent resources
Bottom up: Mitigating attacks from grid edges	Top down: Defending attacks at Control Center or high level
Early warning for anomalies Decentralized approach	Passively defending of emergencies Centralized approach
Harmonizing human activities with the continuing evolution of the community power grids	Grid-centric, business-driven solutions
Empowering smart and connected communities	Defensive solutions – keeping resilient and safe power grid
Integration of data-driven and model-based approaches for self-modeling and autonomy	Designing, planning, and operation based on models

building resilience – that ability to overcome the unexpected and bounce back – in our power grids has become paramount for today's power utilities.

In the foreseeable future, enabled by grid edge technologies, artificial intelligence, and programmable microgrids, power systems will be able to proactively and intelligently anticipate and prepare for emergencies and protect themselves from cyberphysical attacks. These "brain-empowered" smart grids will even be able to adapt to changes in their environments by using the principle of understanding-by-building [23]. In this context, power systems will be not only a built environment but also a human-centered ecosystem that will coevolve with nature's changes and customer activities. This will call for new performance metrics such as *renewability* or *regenerability*. Renewability or regenerability may be different from resilience in the following aspects, as summarized in Table 10.1. As can be seen, the vision of an autonomic and regenerative power system will take years of efforts to be achieved.

References

[1] P. Shah and K. Gehring, "Smart Solutions to Power the 21st Century: Managing Assets Today for a Better Grid Tomorrow," *IEEE Power and Energy Magazine*, vol. 14, no. 2, pp. 64–68, 2016.

[2] F. Noseleit, "Renewable Energy Innovations and Sustainability Transition: How Relevant Are Spatial Spillovers?" *Journal of Regional Science*, vol. 58, no. 1, pp. 259–275, 2018.

[3] E. Bullich-Massagué, F. Díaz-González, M. Aragüés-Peñalba, F. Girbau-Llistuella, P. Olivella-Rosell, and A. Sumper, "Microgrid Clustering Architectures," *Applied Energy*, vol. 212, pp. 340–361, 2018.

[4] Y. Yoldaş, A. Önen, S. Muyeen, A. V. Vasilakos, and I. Alan, "Enhancing Smart Grid with Microgrids: Challenges and Opportunities," *Renewable and Sustainable Energy Reviews*, vol. 72, pp. 205–214, 2017.

[5] M. S. Sadabadi, Q. Shafiee, and A. Karimi, "Plug-and-Play Voltage Stabilization in Inverter-Interfaced Microgrids via a Robust Control Strategy," *IEEE Transactions on Control Systems Technology*, vol. 25, no. 3, pp. 781–791, 2017.

[6] L. Ren, Y. Qin, Y. Li, P. Zhang, B. Wang, P. B. Luh, S. Han, T. Orekan, and T. Gong, "Enabling Resilient Distributed Power Sharing in Networked Microgrids through Software Defined Networking," *Applied Energy*, vol. 210, pp. 1251–1265, 2018.

[7] A. Majzoobi and A. Khodaei, "Application of Microgrids in Supporting Distribution Grid Flexibility," *IEEE Transactions on Power Systems*, vol. 32, no. 5, pp. 3660–3669, 2017.

[8] Z. Shuai, Y. Sun, Z. J. Shen, et al., "Microgrid Stability: Classification and a Review," *Renewable and Sustainable Energy Reviews*, vol. 58, pp. 167–179, 2016.

[9] K. P. Schneider, F. K. Tuffner, M. A. Elizondo, C.-C. Liu, Y. Xu, and D. Ton, "Evaluating the Feasibility to Use Microgrids as a Resiliency Resource," *IEEE Transactions on Smart Grid*, vol. 8, no. 2, pp. 687–696, 2017.

[10] Y. Li, P. Zhang, L. Zhang, and B. Wang, "Active Synchronous Detection of Deception Attacks in Microgrid Control Systems," *IEEE Transactions on Smart Grid*, vol. 8, no. 1, pp. 373–375, 2017.

[11] X. Liu, M. Shahidehpour, Y. Cao, L. Wu, W. Wei, and X. Liu, "Microgrid Risk Analysis Considering the Impact of Cyber Attacks on Solar PV and ESS Control Systems," *IEEE Transactions on Smart Grid*, vol. 8, no. 3, pp. 1330–1339, 2017.

[12] K. Kvaternik, A. Laszka, M. Walker, D. Schmidt, M. Sturm, A. Dubey, "Privacy-Preserving Platform for Transactive Energy Systems," *arXiv preprint arXiv:1709.09597*, 2017.

[13] Q. Shafiee, J. M. Guerrero, and J. C. Vasquez, "Distributed Secondary Control for Islanded Microgrids – a Novel Approach," *IEEE Transactions on Power Electronics*, vol. 29, no. 2, pp. 1018–1031, 2014.

[14] M. A. Allam, A. A. Hamad, M. Kazerani, and E. F. El Saadany, "A Novel Dynamic Power Routing Scheme to Maximize Loadability of Islanded Hybrid AC/DC Microgrids under Unbalanced AC Loading," *IEEE Transactions on Smart Grid*, vol. 9, no. 6, pp. 5798–5809, 2018.

[15] S. Gholami, M. Aldeen, and S. Saha, "Control Strategy for Dispatchable Distributed Energy Resources in islanded Microgrids," *IEEE Transactions on Power Systems*, vol. 33, no. 1, pp. 141–152, 2018.

[16] M. Wolsink, "The Research Agenda on Social Acceptance of Distributed Generation in Smart Grids: Renewable as Common Pool Resources," *Renewable and Sustainable Energy Reviews*, vol. 16, no. 1, pp. 822–835, 2012.

[17] E. M. Gui, M. Diesendorf, and I. MacGill, "Distributed Energy Infrastructure Paradigm: Community Microgrids in a New Institutional Economics Context," *Renewable and Sustainable Energy Reviews*, vol. 72, pp. 1355–1365, 2017.

[18] E. Mengelkamp, J. Gärttner, K. Rock, S. Kessler, L. Orsini, and C. Weinhardt, "Designing Microgrid Energy Markets: A Case Study: The Brooklyn Microgrid," *Applied Energy*, vol. 210, pp. 870–880, 2018.

[19] A. Ali, W. Li, R. Hussain, X. He, B. W. Williams, and A. H. Memon, "Overview of Current Microgrid Policies, Incentives and Barriers in the European Union, United States and China," *Sustainability*, vol. 9, no. 7, p. 1146, 2017.

[20] J. Bongard, V. Zykov, and H. Lipson, "Resilient Machines through Continuous Self-Modeling," *Science*, vol. 314, no. 5802, pp. 1118–1121, 2006.

[21] Z. Bie, P. Zhang, G. Li, B. Hua, M. Meehan, and X. Wang, "Reliability Evaluation of Active Distribution Systems Including Microgrids," *IEEE Transactions on Power Systems*, vol. 27, no. 4, pp. 2342–2350, 2012.

[22] P. Zhang, Y. Wang, W. Xiao, and W. Li, "Reliability Evaluation of Grid-Connected Photovoltaic Power Systems," *IEEE Transactions on Sustainable Energy*, vol. 3, no. 3, pp. 379–389, 2012.

[23] S. Haykin, "Cognitive Radio: Brain-Empowered Wireless Communications," *IEEE Journal on Selected Areas in Communications*, vol. 23, no. 2, pp. 201–220, 2005.

Index